ALGEBRA EXAMPLES

CONIC 5 HYPERBOLAS

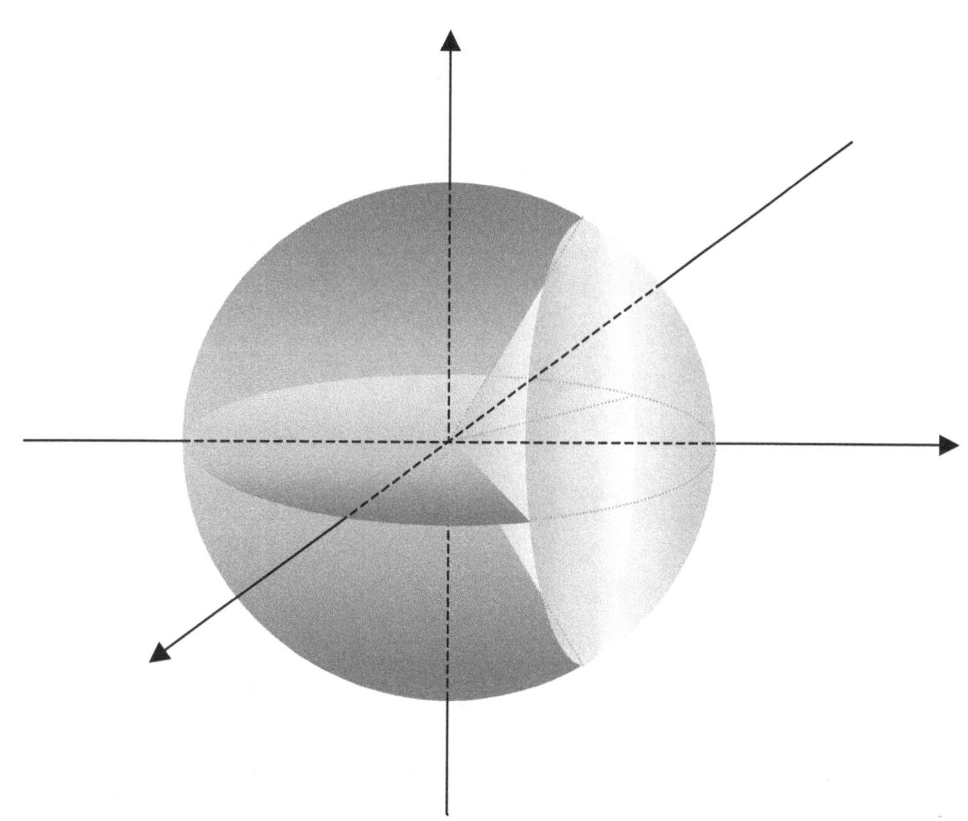

Seong R. Kim

Dear students:

Students need the best teacher, so you need examples, because examples are the best teacher. All the examples here are fully worked, and explain **how** the basic and essential tools in math are made, together with **what** they are, **how** they work, and **how** to work with them. Such tools include numbers, formulas, identities, equations, laws, etc.

Examples here begin with easy ones, of course. Covering every meter and yard properly, we can cover thousands of miles and kilometers. And it is particularly the case in math.

Of those examples therefore, some might even look too easy for you. It's not that easy though, to come up with those examples. Anyways, the bigger and the taller the tree, the deeper and the stronger the root.

Doing math, we work with ideas and run ideas, because every thing in math is an idea. A number is an idea, for instance, and the same is true for a line or circle, too. And putting ideas together, we build another, which becomes the base or an element of another, and each is connected. And that's the way your math grows. So you get to build a circuit, and sometimes, need to fill the gap or repair the circuit so that you get the sense of it.

So your calculation runs properly, and you get the problem solved.

The examples have been made and arranged so that they get tougher (or sometimes easier for some reason) as you proceed with them. In particular, similar examples with some variations are strategically repeated so that you can get the ideas or the tools tricky or complicated, and can get them mastered.

This book is however, nothing but a bunch of examples until you get it powered. How then, to get it powered, and make it run and work for you?

Just read it, and then, do each example in writing. And it is important to note that you do it in **your** writing. Just watching someone doing it, you just only feel that you can do it. If you do it, you can do it, but if you don't, we can hardly. It's a cliché, of course, but is always true that knowing is one thing and doing is another.

I've been helping students grow, take care of, and run their own math. The area covers algebra and geometry for high school or college students, and is especially for equations (for unknowns or curves), functions, and their graphs, which are the basic elements in calculus, which's been the core of my interest from my early age in high school.

Of my students, some are quite poor in math, and thus, are afraid of or hate math, some require special education because of exceptional intelligence, some are smart enough, some are naïve and diligent, some are clever but lazy, and most behave in general. All the students are badly after though, one thing in common: a strong and secure math skill. It is of course, the prime objective of my work, and I'm always happy to and eager to help them achieve it. The problem was however, that many of them wanted it to be purchased. And the question is, can we buy it?

We can buy the means, of course. And a solid math skill is feasible, too. We know however, we can't buy love, and the same is true for the math skill, too. It's not what we can buy or sell, and not what we can give or take. It is however, what we can grow, and need to grow. Your math grows as much as you grow and take care of it. So does mine.

What math then, do students most often do or use in high schools or colleges?

It is algebra and geometry. What algebra though?

Elementary algebra, of course
Doing the algebra, we work with numbers (many in kinds), constants, variables, ratios, rates, expressions, equations, inequalities, functions, identities, formulas, laws, etc., together with signs and symbols. And if we want to do algebra properly, we want to know their natures and how they mingle with each other.

So studying math ideas or tools, you want to know **what** they are, **how** they work, and **how** to work with them or **what** to do with them.   What then, about the geometry?

Basically, the geometry has much to do with shapes, positions, and angles. The shapes begin with triangles and circles, and move on to rectangles, squares, parallelograms or rhombuses, trapezoids, tetragons, other polygons, polyhedrons, etc.

Doing the geometry, too, though, we need to do the algebra stated above. So it is analytic geometry, often called coordinate geometry, too. And doing it, we can specify positions using coordinates. So in the geometry, basically, we work with graphs. Putting a math idea in a graph, we can not only effectively think about it but actually see it, too, and therefore, can efficiently work with it. What idea then, is it?

The idea begins with a point, line, parabola, circle, ellipse, and hyperbola, called a conic section or basic curve, and then, moves on to other curves, planes, surfaces, volumes, and other objects in various dimensional spaces, together with vectors.

And using an angle, we can specify an amount of turn or change in direction.

So learning, using, or applying those ideas or math tools, we get to solve problems.

And this book can help. It can help learn them, and use them so that you can navigate to find solutions to problems. And in particular, it can help come up with answers to those **what**s and **how**s stated above. So it can help you grow and run your own math, and thus, can help achieve your solid math skill.

It is however, not a magic book giving you a math skill of high caliber overnight. And it can have many mistakes, too. There is no magic, and math is full of facts and ideas. And it is after all, not me and not your teacher but you who put together some of those facts and ideas, and understand it. Putting facts and ideas together, understanding it, and taking care of what you have learned, you grow your math. And this book can help.

This is a book of examples designed to help you grow your math, and assumes that you are a real beginner. This book requires though, time and effort, the amount of which need to be substantial, too, but will be worth it. That's because you want a substantial achievement, and will get it. And probably, you will get to see this book helping you get there much faster than expected. And then, you will get to see the way math runs.

In math, everything is an idea. So is a problem. And solving it, we put it many different ways. For instance, while expanding or reducing it, or modifying or converting it, we keep searching for the solution, approaching the solution, and eventually, can get there. So don't look for the solution outside the problem. The solution is inside the problem if the problem is properly made.

If it is not, no solution is the solution. And in fact, it is often the case a problem itself is the solution. We can put a problem in many different ways, and eventually, can end up with the solution.   How come then, is the solution no other than the problem?

For instance, the solution to $3232 \div 101$ is 32.   And we can put it this way:

$$3232 \div 101 = \frac{3232}{101} = \frac{32 \times 101}{101} = \frac{32}{1} = 32 \implies 3232 \div 101 = 32.$$

And we can get this, too: $32 \implies 3232 \div 101$.   How?

$$32 = \frac{32}{1} = \frac{32 \times 101}{101} = \frac{3232}{101} = 3232/101 = 3232 \div 101.$$ Too easy?

For another instance, the solution to $ax^2 + bx + c = 0$ is: $x = \frac{-b \pm \sqrt{b^2 - 4ac}}{2a}$, which is called the quadratic formula.   How come then, is the solution no other than the problem?

We can put it this way:

$$x = \frac{-b \pm \sqrt{b^2 - 4ac}}{2a} \implies 2ax = -b \pm \sqrt{b^2 - 4ac} \implies 2ax + b = \pm\sqrt{b^2 - 4ac}$$

$$\implies (2ax + b)^2 = b^2 - 4ac \implies 4a^2x^2 + 4abx + b^2 = b^2 - 4ac$$

$$\implies 4a^2x^2 + 4abx = -4ac \implies ax^2 + bx = -c \implies ax^2 + bx + c = 0.$$

And we can get this, too: $ax^2 + bx + c = 0 \implies x = \frac{-b \pm \sqrt{b^2 - 4ac}}{2a}$.   How?

$$ax^2 + bx + c = a(x^2 + \tfrac{b}{a}x) + c = a(x^2 + \tfrac{b}{a}x + \tfrac{b^2}{4a^2} - \tfrac{b^2}{4a^2}) + c = a(x^2 + \tfrac{b}{a}x + \tfrac{b^2}{4a^2}) - \tfrac{b^2}{4a} + c$$

$$= a(x + \tfrac{b}{2a})^2 - \tfrac{b^2 - 4ac}{4a} = 0 \implies a(x + \tfrac{b}{2a})^2 = \tfrac{b^2 - 4ac}{4a} \implies (x + \tfrac{b}{2a})^2 = \tfrac{b^2 - 4ac}{4a^2} \implies x + \tfrac{b}{2a} = \pm\sqrt{\tfrac{b^2 - 4ac}{4a^2}}$$

$$\implies x = -\tfrac{b}{2a} \pm \tfrac{\sqrt{b^2 - 4ac}}{2a} = \tfrac{-b \pm \sqrt{b^2 - 4ac}}{2a} \implies x = \tfrac{-b \pm \sqrt{b^2 - 4ac}}{2a}.$$

And we call the set of processes above, algebra.

So if a problem is well defined, that is, if it makes sense, we should be able to get it solved the way below:

A problem ⇒ ... ⇒ ... ⇒ the solution, and thus: **the problem ⇒ the solution**.

So solving a problem, we put it many different ways so that we can get to the solution.

And that's the way, math runs.

May your math run very well.

Seong R. Kim

B.S. Math. Michigan Tech. Univ.   M.S. Math. Rensselaer Polytechnic Institute

Notes:

This book is one of five books about some basics in elementary algebra, and covers equations often used in high schools and colleges or universities. And the equations are for curves called conic sections, often just called conics. There are five kinds in conics. And of the five, one is covered briefly here in this book, and is for hyperbolas.

So this book covers equations indicating hyperbolas, that is, equations for hyperbolas. And this book explains what such an equation is about, how it gets made, what it does or how it behaves, and what we can do with it or how to use it.   What then, is it for?

A hyperbola is an idea in math, so it's a math idea, and is a tool in math. So it's a math tool. And we use it, solving problems, of course. So students need to get the idea.

And thus, this book helps you get the idea of a hyperbola, that is, the concept of a math object called a hyperbola, and see how to use it, because the book explains what it is and how it works, along with those stated above so that you can develop your own idea to make use of it, solving problems, of course.   What then, about the other conics?

They are covered in their individual books, too, which are as follows:

Algebra Examples Conics 1 Lines, which covers therefore, equations for lines, often called linear equations or equations of degree 1.

Algebra Examples Conics 2 Parabolas, which covers thus, equations for parabolas, often called quadratic equations or equations of degree 2.

Algebra Examples Conics 3 Circles, which is about equations for circles.

Algebra Examples Conics 4 Ellipses, explaining those for ellipses.

And each book is designed for those students who want to study calculus, want to major in science or engineering, or want to take IB courses in math, so each book covers materials in each category in such a depth and extent. And if you don't need that much, it will be sufficient to study the sections and the sets of examples bulleted in the table of contents.

Either way, the books will help you grab math ideas often used in real life as well as in math courses. The ideas are about lines, parabolas, circles, ellipses, hyperbolas, and their equations, so you will get to see what those curves and equations are about, how they work, and how to use them, and develop your own idea to make use of those, solving problems, of course.

In short, the books help you develop and grow your own idea to make use of math ideas, providing examples, showing all the steps and the ideas behind them, and explaining what the math ideas are about.

Contents

$(x + y)^2 = x^2 + 2xy + y^2.$ $(x + y)^3 = x^3 + 3x^2y + 3xy^2 + y^3.$

$(x + y)(x - y) = x^2 - y^2.$ $(x + y)(x^2 - xy + y^2) = x^3 + y^3.$

$(x^2 + xy + y^2)(x^2 - xy + y^2) = x^4 + x^2y^2 + y^4.$

$(x + a)(x + b) = x^2 + (a + b)x + ab.$ $(ax + b)(cx + d) = acx^2 + (ad + bc)x + bd.$

$(x + a)(x + b)(x + c) = x^3 + (a + b + c)x^2 + (ac + bc + ca)x + abc.$

$(a + b + c)^2 = a^2 + b^2 + c^2 + 2(ab + bc + ca).$

$(a + b + c)(a^2 + b^2 + c^2 - ab - bc - ca) = a^3 + b^3 + c^3 - 3abc.$

Suppose both a and $b \neq 0$, and both m and n are integers. Then, we get:

0. $a^m a^n = a^{m+n}$ **1.** $a^m / a^n = \dfrac{a^m}{a^n} = a^{m-n}$ **2.** $(a^m)^n = a^{mn}$

3. $(ab)^n = a^n b^n$ **4.** $(a/b)^n = \left(\dfrac{a}{b}\right)^n = a^n / b^n = \dfrac{a^n}{b^n}$

Suppose both a and $b > 0$, and m and n both are integers nonzero. Then, we get:

0.1. $a^{\frac{1}{n}} b^{\frac{1}{n}} = (ab)^{\frac{1}{n}}.$ **1.1.** $\dfrac{a^{\frac{1}{n}}}{b^{\frac{1}{n}}} = \left(\dfrac{a}{b}\right)^{\frac{1}{n}}.$ **2.1.** $(a^{\frac{1}{n}})^m = (a^m)^{\frac{1}{n}}.$

3.1. $(a^{\frac{1}{n}})^{\frac{1}{m}} = a^{\frac{1}{mn}} = (a^{\frac{1}{m}})^{\frac{1}{n}}.$ **3.2.** $(a^{mp})^{\frac{1}{np}} = (a^m)^{\frac{1}{n}}$, where p is a nonzero integer.

1. Suppose M, N, and $b > 0$, but $b \neq 1$, and we have: $A = \log_b M$, and $B = \log_b N$. Then, we get: $A - B = \log_b M - \log_b N = \log_b \frac{M}{N}$.

2. Suppose that M and $b > 0$, but $b \neq 1$, and that we have: $E = \log_b M$. Then, we get: $PE = P \log_b M = \log_b M^P$.

3. Suppose that a, b, C, and $D > 0$, but a and $b \neq 1$, and that we have: $\log_a C = \log_b D$. Then, we get: $\log_a C = \log_b D = \log_{ab} CD$.

4. Suppose that a, b, C, and $D > 0$, but a and $b \neq 1$, and that we have: $\log_a C = \log_b D$. Then, we get: $\log_a C = \log_b D = \log_{\frac{a}{b}} \frac{C}{D} = \log_{\frac{b}{a}} \frac{D}{C}$.

5. $\log_b b = 1$, and $\log_b 1 = 0$.   **6.** $\log_b A = \dfrac{\log_c A}{\log_c b}$.

7. $\log_b A = \dfrac{1}{\log_A b}$.

Note:

The drawings or graphs in this book are not exact, and are approximate or conceptual ones.

$\in$	"$a \in B$" means that a belongs to B. "$p, q,$ **and** $r \in W$" means that $p, q,$ and r belong to W.
$\Rightarrow$	"$A \Rightarrow B$." means that A implies B.
$\equiv$	$A \equiv B$ means that A and B are identical to each other.
$\neq$	$A \neq B$ means that A is not equal to B.
$\lvert A \rvert$	The magnitude of A. For instance, $\lvert -1 \rvert = \lvert 1 \rvert = 1$.
$\therefore$	Therefore
$\Leftrightarrow$	"$A \Leftrightarrow B$" means "If A then B." and "If B then A." We can read $A \Leftrightarrow B$ as "A if and only if B." In such a case, we can say that $A = B$.
Δx and Δy	Suppose that (x_1, y_1) and (x_2, y_2) are two points in the x-y plane. Then, we get either of the two below. $\Delta x = x_2 - x_1$, and $\Delta y = y_2 - y_1$. $\Delta x = x_1 - x_2$, and $\Delta y = y_1 - y_2$.

Distance Formula

Suppose that d is the distance between two points (x_1, y_1) and (x_2, y_2) in the x-y plane. Then, we get $d^2 = (\Delta x)^2 + (\Delta y)^2$.

₀. **What is a hyperbola?**

Like an ellipse, a hyperbola is a conic section, often just called a conic, too.

So it is a conic, and is a basic curve. Unlike other conics though, it is not just a single curve, but is made of two. Each is called a branch (or arm). So a hyperbola is made of two branches. And a branch looks like a parabola, yet is quite different from a parabola.

Though look is quite different, a hyperbola and an ellipse have quite a few in common.

A hyperbola has a center and two axes of symmetry, called the main axes, too. One is a major axis, and the other is a minor axis, called an imaginary axis, too. And the two axes are perpendicular to each other, and meet at the center. The <u>major</u> axis is however, often called the <u>transverse</u> axis, and the <u>minor</u> axis is often called the <u>conjugate</u> axis.

Next, like an ellipse, a hyperbola has two foci, too, each of which has a corresponding directrix. Thus, a hyperbola has two directrices, too, and each branch has a focus and a directrix corresponding to the focus.

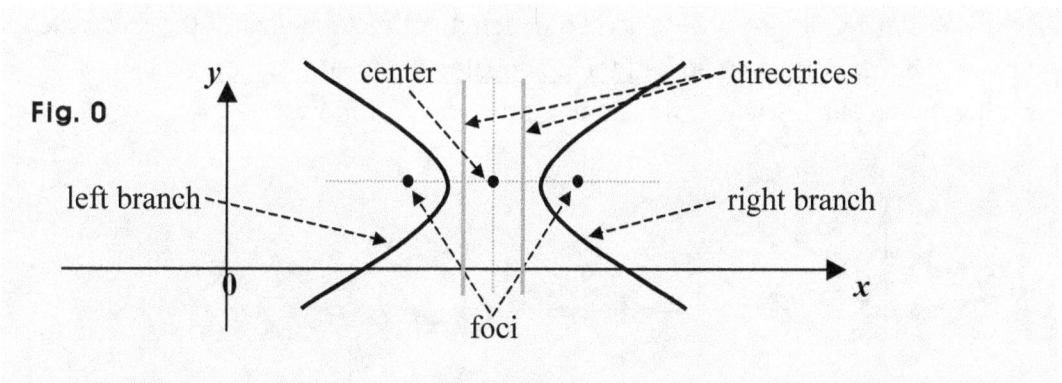

Fig. 0

So we can say that a hyperbola is made of two braches that look like parabolas.

In the figure below, the two dotted lines are perpendicular to each other, and meet at the center. And the dotted line <u>passing through the foci</u> is called the <u>transverse line</u>. And the other line dotted is called the <u>perpendicular bisector</u> of the transverse line.

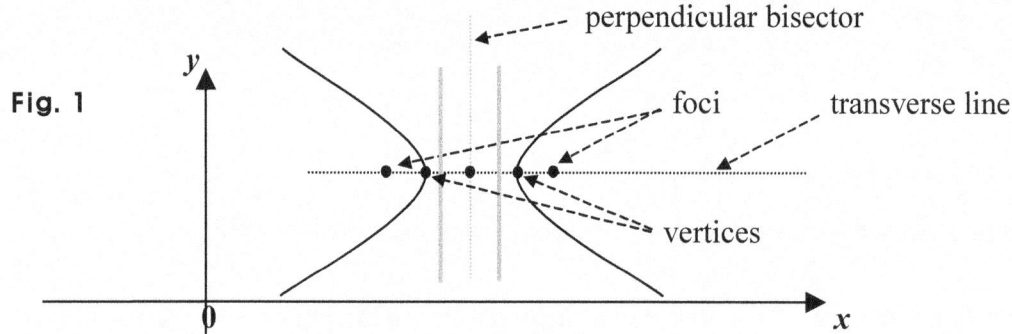

Fig. 1

Also, like an ellipse, a hyperbola has two vertices. Each branch has one vertex. And the <u>two vertices</u> are the <u>endpoints of the transverse axis</u>, and thus, are in the transverse line.

So the length of <u>the transverse (major) axis</u> is the distance <u>between the two vertices</u>, and <u>the transverse line</u> has: <u>the vertices, together with the foci and the center</u>.

Next, a hyperbola has a focal distance, too, which is, of course, the distance from either focus to <u>the center</u>, which is thus, <u>the midpoint between the foci</u>.

Where then, are the two main axes, that is, the transverse axis and the conjugate axis?

A hyperbola is said to have an invisible rectangle.
And it can be a square, too. Anyway, taking the length of the diagonal of the rectangle, we get twice the focal distance. So <u>half the diagonal is the focal distance</u>, too.
And making the rectangle visible, we can put it the way as follows:

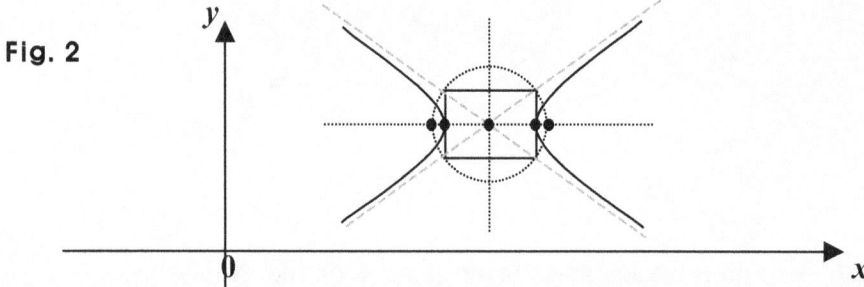

Fig. 2

We can see in the figure, the rectangle is tangent to both branches at the vertices of the hyperbola, and its <u>diagonal</u> is the <u>diameter</u> of the circle that passes through all the vertices of the rectangle. The circle is said to be circumscribed about the rectangle. And such a circle is called a circumcircle. So the diagonal is the diameter of the circumcircle.

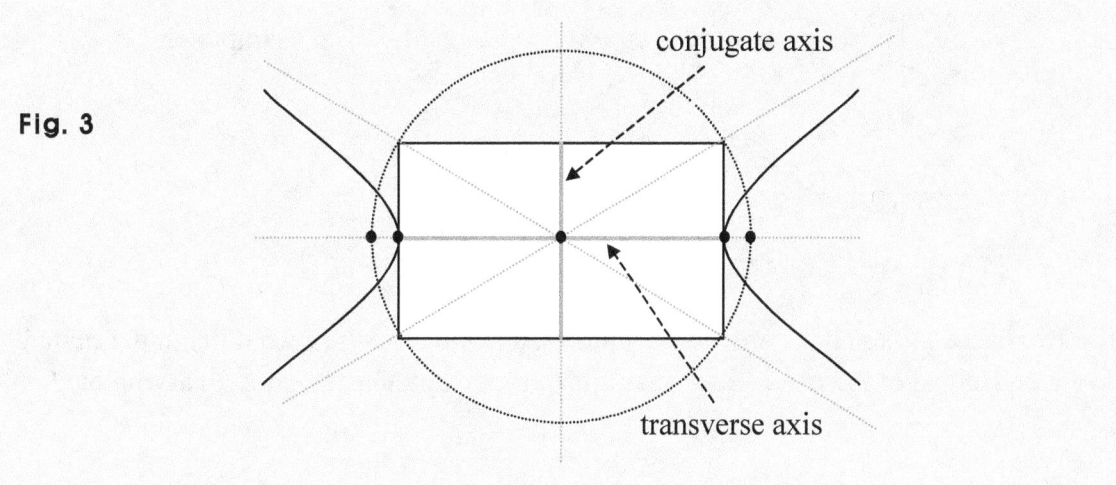

Fig. 3

conjugate axis

transverse axis

And the diameter is the distance between the two foci.
So half the diameter, that is, <u>half the diagonal is the focal distance</u>.
In other words, <u>the radius</u> is the focal distance.

Then, the <u>transverse axis</u> is the line segment connecting the vertices of the hyperbola, that is, the <u>length of the rectangle</u>. And the <u>conjugate axis is the width</u> of the rectangle.

Also, <u>half the length of the rectangle</u> is called the <u>semi major axis</u>, and <u>half the width</u> is called the <u>semi minor axis</u>.

In short, <u>half the transverse</u> is the <u>semi major</u>, and <u>half the conjugate</u> is the <u>semi minor</u>.

So the rectangle has most of the information (specifications) of the hyperbola that has the rectangle.
And thus, specifying the dimensions of the rectangle, together with the center, we can describe specifically (that is, define) the hyperbola.

So understanding and using the rectangle is important.

Next, unlike other conics, a hyperbola has two special lines, called <u>asymptotes</u>.

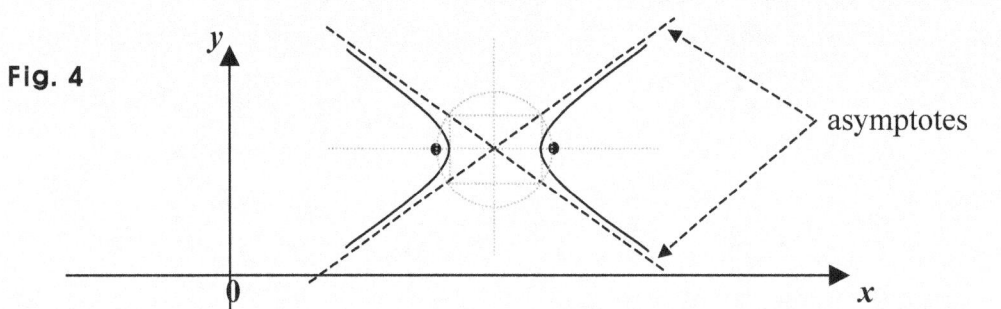

Fig. 4

asymptotes

The two dashed lines above are the asymptotes, and are crossing each other at the center. So each diagonal of the rectangle is a part of each asymptote.   What is an asymptote?

Suppose a point is moving along a branch of a hyperbola.

Then, there is a line that the point approaches as the point gets farther away from the vertex. No matter how far the point may be however, from the vertex, the point will never meet the line. And we call the line an asymptote, which a dashed line above.

So as the point gets farther and farther away from the vertex, the distance between the point and the asymptote gets smaller and smaller, but won't be 0. In other words, the distance can be as good as 0, but cannot be 0. So the point doesn't meet the asymptote.

And a hyperbola has two asymptotes, and the two meet at the center, but not necessarily at a right angle, 90°. Thus, the asymptotes do not have to be perpendicular to each other. So they can meet at an angle other than 90°.

And as in the case of ellipses, **in this book**, if the asymptotes are perpendicular to each other, the hyperbola is said to be rectangular. So it is called a rectangular hyperbola. <u>Note however, in other books, it may not be called that way</u>.

If a hyperbola is rectangular, the invisible rectangle is a square, because the asymptotes are perpendicular to each other.

And also, **in this book**, <u>if the transverse axis is horizontal</u>, that is, parallel to the *x*-axis, it is called a <u>horizontal hyperbola</u>, and if the axis is vertical, that is, parallel to the *y*-axis, it is called a vertical hyperbola. In other books though, it may not be called that way.

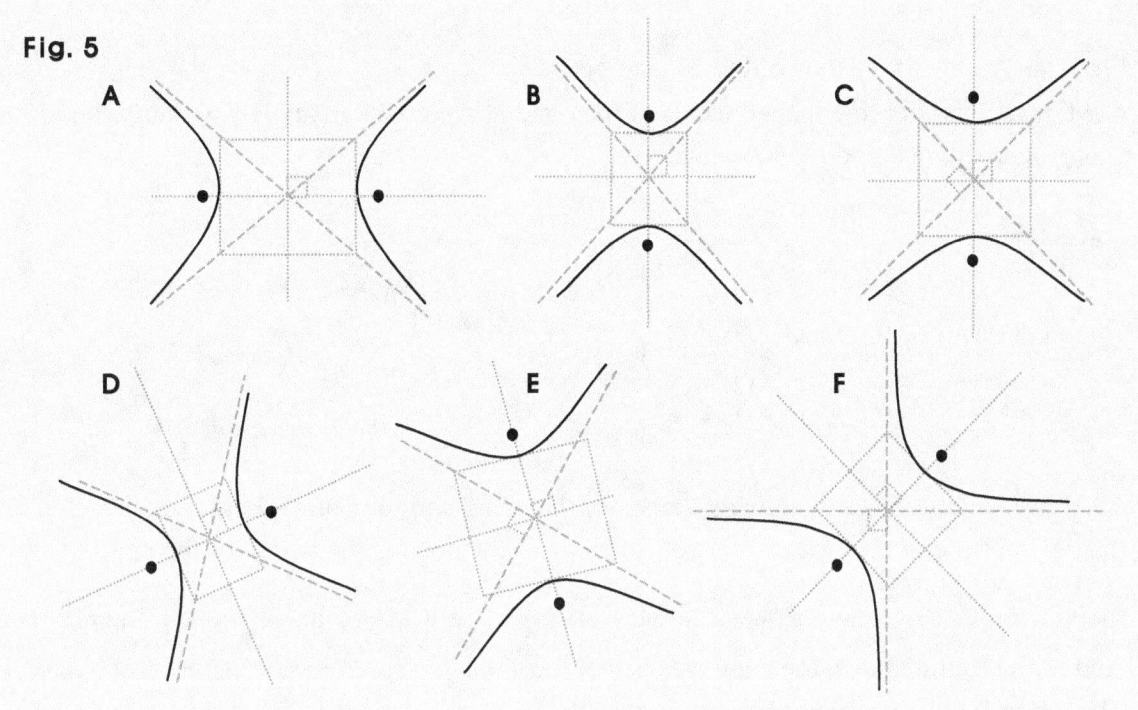

Fig. 5

In the figure above, **A** is a horizontal hyperbola, **B** is a vertical hyperbola, **C** is a vertical rectangular hyperbola, and **E** and **F** are rectangular hyperbolas. And **D** is just called a nonstandard hyperbola. If a hyperbola is standard, it is either horizontal or vertical. So **F** is a nonstandard hyperbola, too. And **A**, **B**, and **C** are standard hyperbolas.

In high school math or basic courses in university math, we usually work with standard ones as **A**, **B**, and **C** above, or if it is nonstandard, it is probably a rectangular hyperbola where asymptotes are parallel or perpendicular to the coordinate axes as **F** above. So in high school math, we don't normally use nonstandard hyperbolas as **D** and **E** above.

How then, can we get a hyperbola?

As shown in the figure below, if cutting a right cone with a plane, we can get a cross section called a hyperbola. So it's called a conic section, just called a conic, for short.

A right cone is a cone, where the line connecting the vertex of the cone and the center of the base is perpendicular to the base, and thus, makes a right angle (90°) with the base.

And the line stated above is called the axis of the cone. And in the study of conics, such a right cone is said to be made or generated the way as follows:

First, take a line as the axis of the cone to be made.
Next, make another line meet the axis of the cone at a point at an angle **γ**, which is an acute angle, that is, **0 < γ < 90°**.

Fig. 6

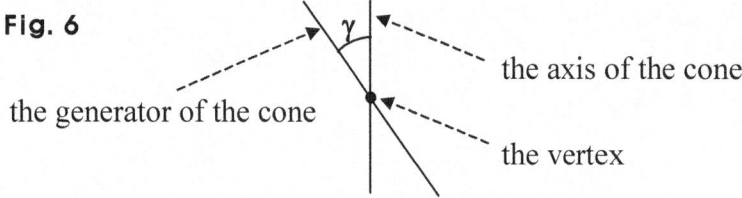

the axis of the cone

the generator of the cone

the vertex

Then, we call the other line the *generator of the cone*, and the point where the other line, that is, the generator of the cone meets the axis of the cone is the vertex of the cone.

So the vertex of the cone is the point where the generator meets the axis of the cone. And in the figure above, the angle **γ** is called the *generator angle*.
Then, rotating the generator around the axis fixing the generator at the vertex, we get a double-cone as shown below:

Fig. 7

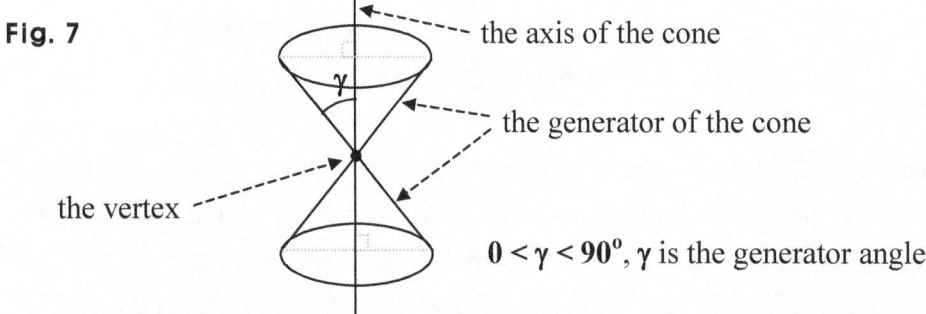

the axis of the cone

the generator of the cone

the vertex

0 < γ < 90°, γ is the generator angle

And note that the generator and the axis are lines, and lines have infinite lengths, so the lengths of the cones are infinite, too. So we cannot show them all. Showing thus, such a cone or a curve in math, we just show some part of it.
How then, can we get a hyperbola cutting the cone with a plane?

The cross section in black below is a pair of curves, and is a hyperbola. So if α is the angle between the plane and the axis of the cone, and $0 \leq \alpha < \gamma$, and if the plane does not include the vertex, the cross section is a pair of curves, and is a conic called a <u>hyperbola</u>.

Fig. 8

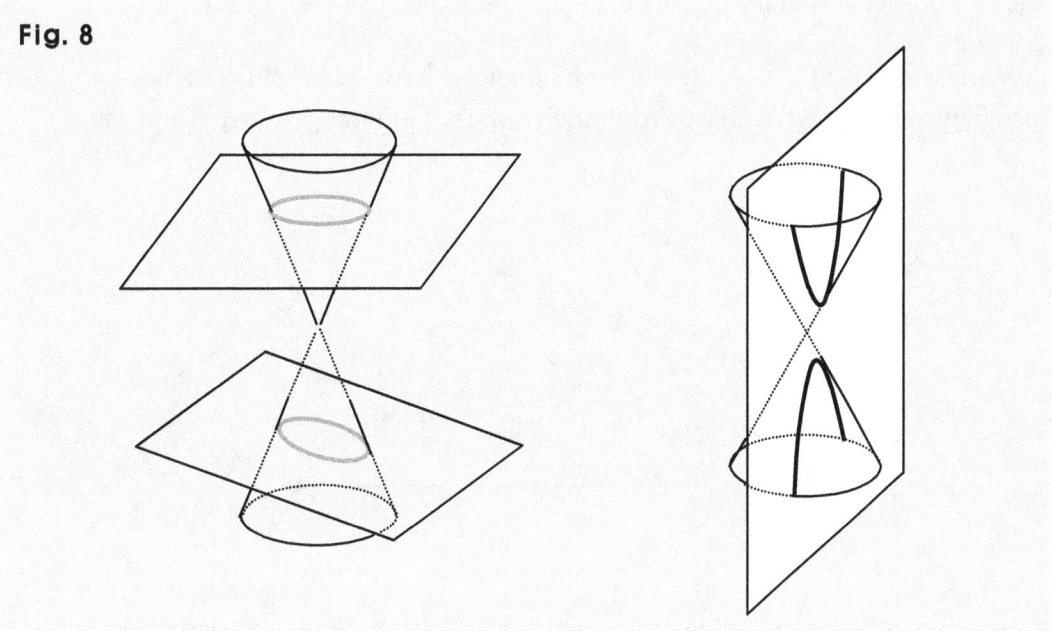

And a sketchy definition of a hyperbola can be as follows:

Suppose we designate two particular points in the *x-y* plane as shown in the figure below.

Then, a hyperbola is a set of points, from each of which, <u>the difference between the two distances to the two particular points is constant</u>, that is, the same. And the designated two points are called the foci of the hyperbola.

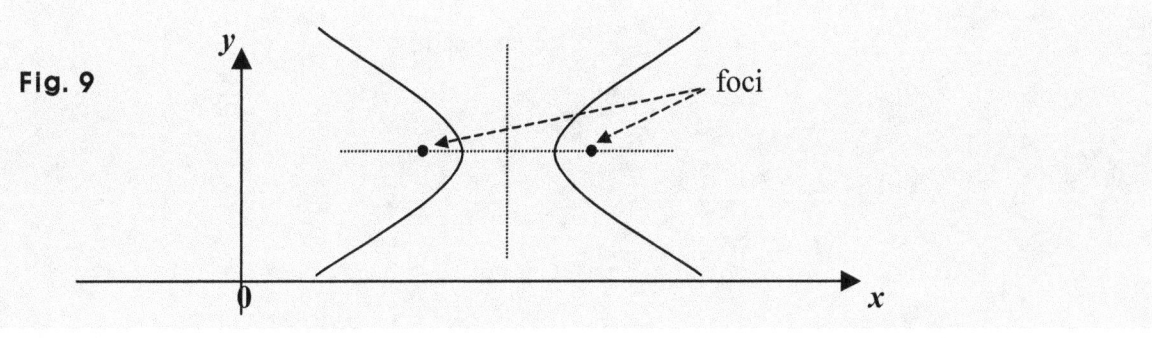

Fig. 9

And also, we can describe (define) a hyperbola the way as follows, too:

Suppose a point is moving along a curve in the *x-y* plane, and the difference between the two distances from the moving point to two particular points is constant.
Then, the two particular points are called the foci, and the curve is a hyperbola.

If therefore, a point (*x, y*) is in a hyperbola, no matter where the point (*x, y*) may be, the difference between the two distances from the point (*x, y*) to the two foci is constant.

Fig. A

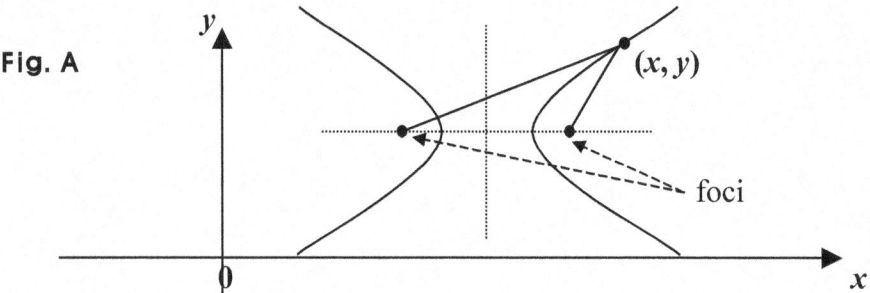

So using the fact above, we can actually make a hyperbola the way as follows.

First, set the two foci and another point by fastening three nails or thumb tacks to a panel as shown below. Then, make a loop using a string and attach to the loop a spring coil or a rubber band. Then, put the loop on the panel so that it wraps around the nails as below.

Fig. B

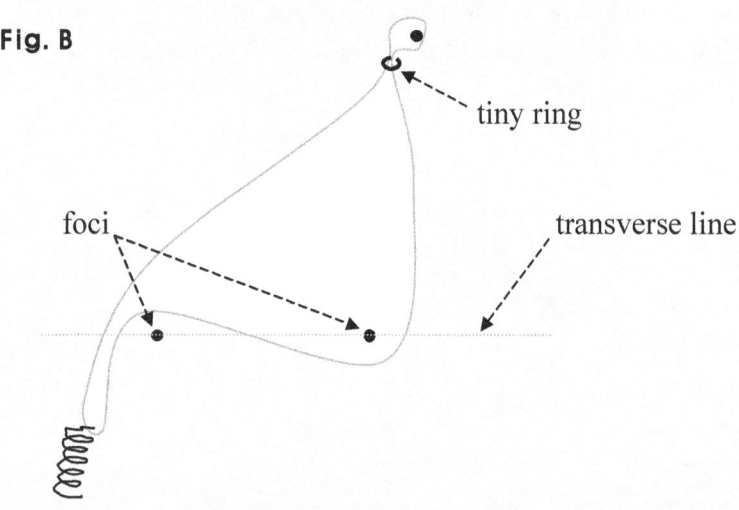

Fig. C

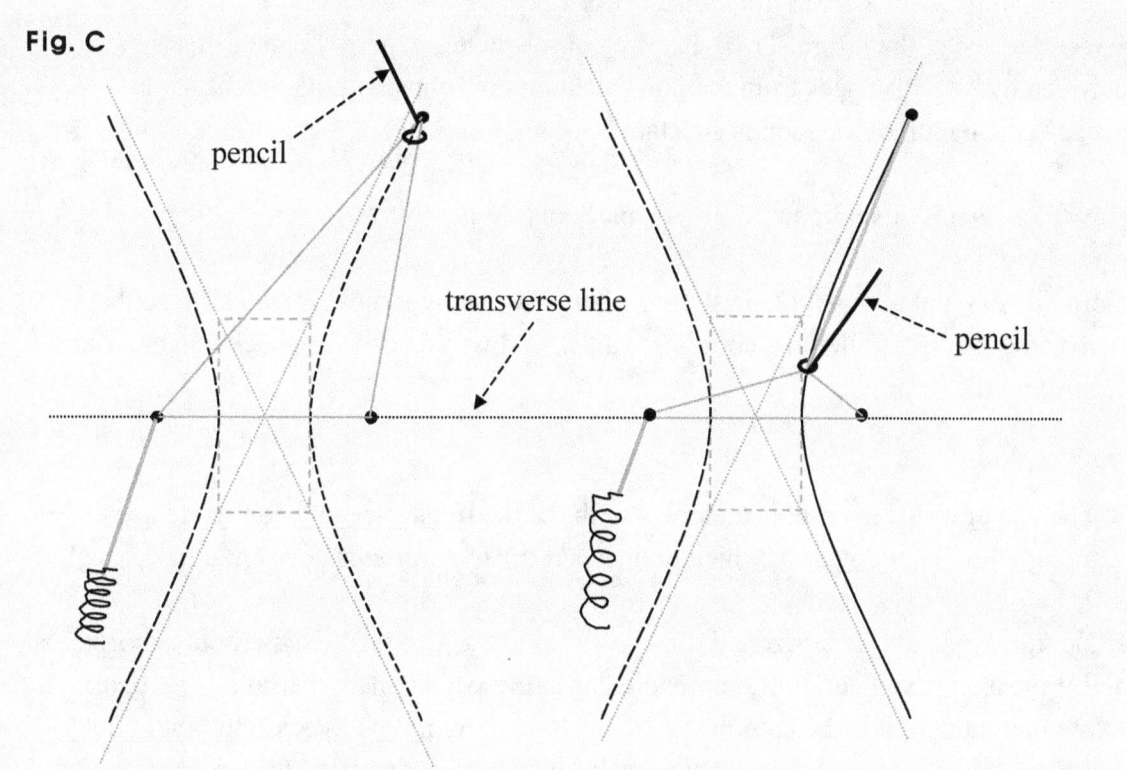

pencil

transverse line

pencil

And as shown below, changing the position of the beginning point, we can change the location of the vertex and the bending in the branch. The more it bends, the closer the vertex is to the focus.

Fig. D

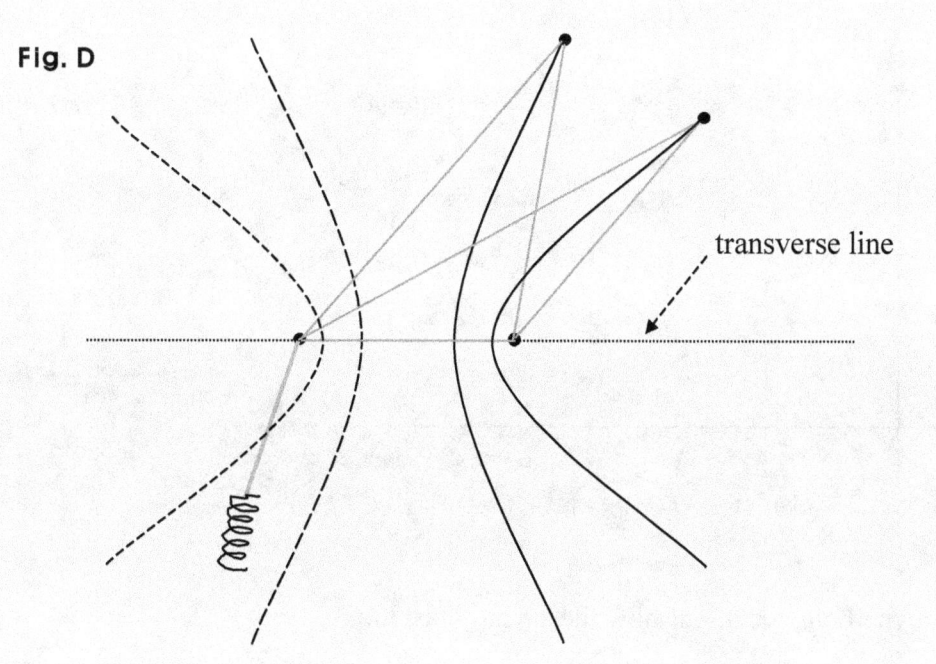

transverse line

So we can see in the figure above that if a point is in the hyperbola, the difference between the two distances from the point to the foci is constant, that is, the same, because the length of the loop is constant.

How then, can we explain or describe a particular hyperbola?

Normally, we put a hyperbola in the *x-y* plain, and the hyperbola is standard, so the transverse axis is parallel to a coordinate axis. And we can describe such a hyperbola specifying the information below:

The <u>center</u> of the hyperbola and the lengths of the <u>transverse and conjugate axes</u>, along with the <u>orientation</u>, which indicates if the hyperbola is horizontal or vertical

If the transverse axis is horizontal, thus, parallel to the *x*-axis, the hyperbola is horizontal, and if the axis is vertical, thus, perpendicular to the *x*-axis, the hyperbola is vertical. (Note that though, it is the case in this book, and <u>it may not be the case in other books</u>. That is to say that in other books, hyperbolas may not be classified the way above, and thus, may not be said to be horizontal or vertical.)

Anyway, for instance, if the center is (5, 2), and the transverse axe is 6, and is parallel to the *x*-axis, and the conjugate axis is 4, the hyperbola is horizontal, and is as follows:

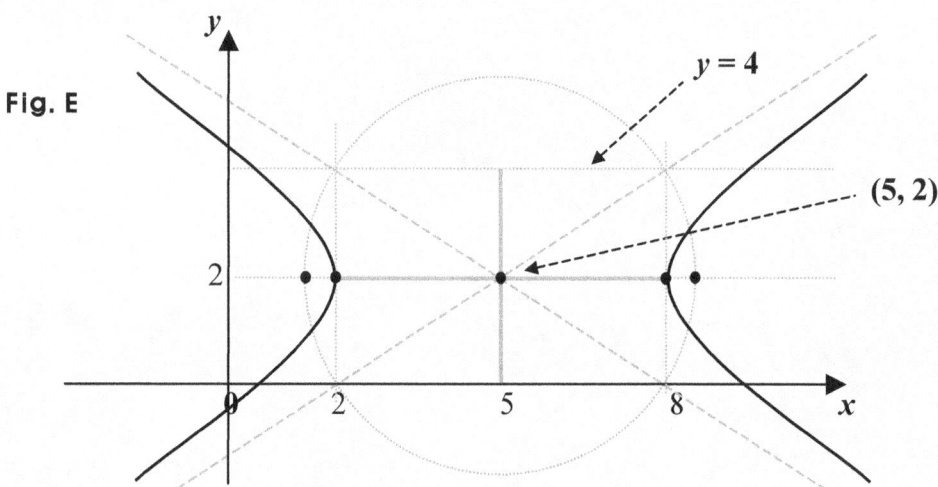

Fig. E

And showing a hyperbola, we often show the asymptotes, too.

And in fact, if specifying the dimensions of the rectangle, that is, the length and the width, along with the center and orientation, we can recognize a particular hyperbola.

So if saying that the rectangle is 6 by 4, centered at (5, 2), and <u>horizontal</u>, we mean the hyperbola shown above.
And if saying that the rectangle is 4 by 6, centered at (5, 2), and <u>vertical</u>, we mean the hyperbola shown below.

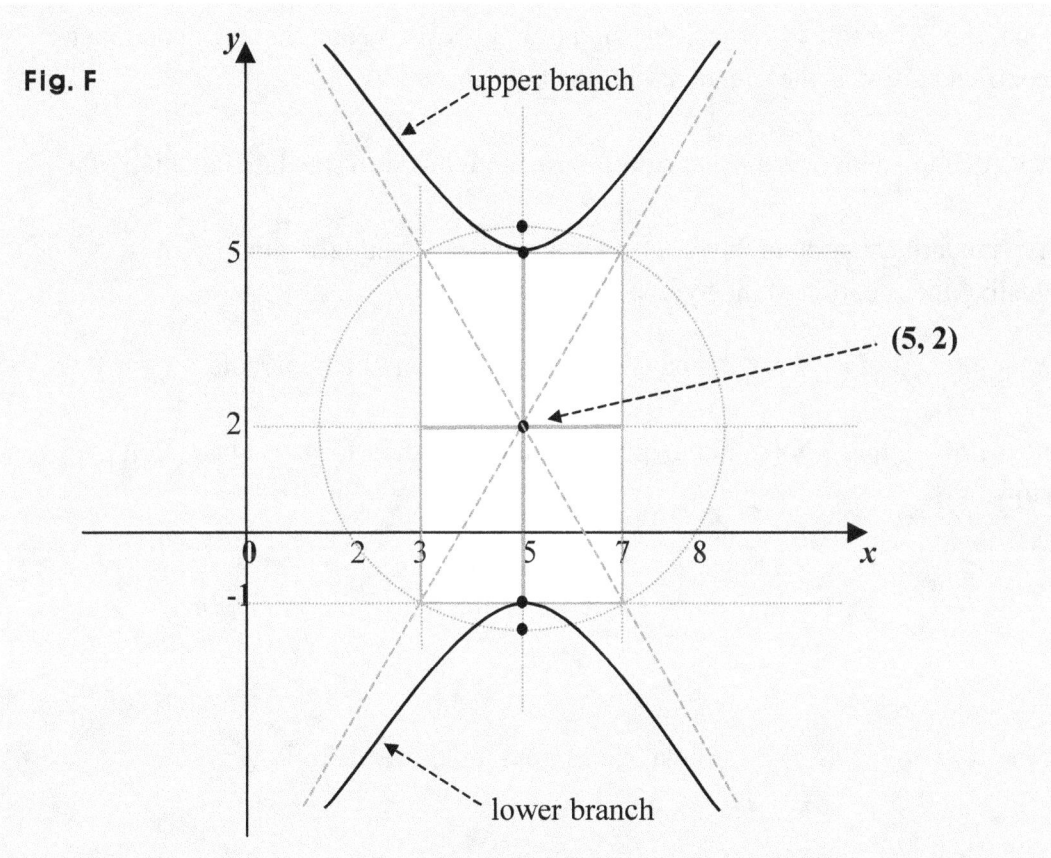

Fig. F

upper branch

(5, 2)

lower branch

In math though, making a hyperbola, we define it. So defining a hyperbola particular, we make the particular hyperbola. And defining a hyperbola, we use an equation indicating the hyperbola. So we can define the particular hyperbola producing the equation of it.

Assuming thus, **H** is the <u>horizontal</u> hyperbola above, we can define **H** the way below:

$$\frac{(x-5)^2}{3^2} - \frac{(y-2)^2}{2^2} = 1.$$

Notice that if a hyperbola is <u>horizontal</u>, <u>the sign of x^2 is positive</u>, and that the denominator of the coefficient of x^2 is the square of half the transverse axis.

And assuming V is the <u>vertical</u> hyperbola above, we can define V the way as follows:

$$\frac{(y-2)^2}{3^2} - \frac{(x-5)^2}{2^2} = 1.$$

Notice that if a hyperbola is <u>vertical</u>, <u>the sign of y^2 is positive</u>, and that the denominator of the coefficient of y^2 is the square of half the transverse axis.

And each equation above is in the standard form, and thus, is a standard equation of a hyperbola.
Putting a standard equation in the general form, we just expand (or simplify) it.
So expanding the equation of the hyperbola H, we get:

$4x^2 - 9y^2 - 40x + 36y + 28 = 0$, which is a general equation of a hyperbola.

And in general, putting a horizontal hyperbola in the standard form, we can put it the way below:

$$\frac{(x-u)^2}{a^2} - \frac{(y-v)^2}{b^2} = 1, \text{ where } a \text{ and } b > 0. \text{ Then, the transverse axis is } 2a.$$

And in the case of a vertical hyperbola, we can put it the way as follows:

$$\frac{(y-v)^2}{b^2} - \frac{(x-u)^2}{a^2} = 1, \text{ where } a \text{ and } b > 0. \text{ Then, the transverse axis is } 2b.$$

So if we define a particular hyperbola placed in the x-y plane, we want to specify the center and the orientation, together with the transverse and conjugate axes (or the semis).

What then, is the center?

If the equation is either of the two above, the center of the hyperbola is a point (u, v).

Note that though, <u>the standard equations</u> above can indicate hyperbolas <u>standard only</u>.

Is there any equation then, that can indicate a hyperbola standard or not?

Putting a hyperbola in such an equation, we can put it the way below:

$$ax^2 + by^2 + cxy + ux + vy + w = 0, \text{ where } ab < 0, \text{ and } 4ab < c^2.$$

Note that *ab* < 0 is saying that *a* and *b* have <u>opposite signs</u>. And of course, in the equation above, *a*, *b*, *c*, *u*, *v*, and *w* are constant. Note <u>however, for some values of the constants, the equation does not indicate a hyperbola, even if *ab* < 0, and 4*ab* < *c*²</u>.

What then, about an equation that is in the <u>general form</u> and can indicate a hyperbola <u>standard only</u>?

Putting a hyperbola in such an equation, we can put it the way below:

$$ax^2 + by^2 + ux + vy + w = 0, \text{ where } ab < 0.$$

So the equation above indicates a hyperbola with the transverse axis parallel to a coordinate axis. And notice that it does not have the *xy*-term, that is, *cxy* in the equation shown earlier. Note <u>however, for some values of the constants, the equation does not indicate a hyperbola, even if *ab* < 0</u>.

And we can give a name to an equation, too. For instance, assuming *H* is the equation above, we can put *H* the way as follows:

$$H(x, y) = ax^2 + by^2 + ux + vy + w = 0, \text{ where } ab < 0.$$

What then, about the standard equation?

Assuming S is the equation $\dfrac{(x-u)^2}{a^2} - \dfrac{(y-v)^2}{b^2} = 1$, we can put S the way as follows:

$$S(x,y) = \dfrac{(x-u)^2}{a^2} - \dfrac{(y-v)^2}{b^2} - 1 = 0, \text{ where } a \text{ and } b > 0.$$

And assuming U is the equation $\dfrac{(y-v)^2}{b^2} - \dfrac{(x-u)^2}{a^2} = 1$, we can put U the way below:

$$U(x,y) = \dfrac{(y-v)^2}{b^2} - \dfrac{(x-u)^2}{a^2} - 1 = 0, \text{ where } a \text{ and } b > 0.$$

What if $a = b$ though, in the equation above?

Then, the equation still indicates a hyperbola.   What hyperbola then, is it?

It is the hyperbola where the both main axes are a (or b), and the center is at (u, v).

Setting: $b = a$ in S, we get: $\dfrac{(x-u)^2}{a^2} - \dfrac{(y-v)^2}{b^2} - 1 = 0 \Rightarrow \dfrac{(x-u)^2}{a^2} - \dfrac{(y-v)^2}{a^2} = 1$

So the equation indicates a hyperbola where the transverse axis is **2a**, which is also, the conjugate axis, and the center is (u, v). And of course, the hyperbola is horizontal.

And notice that it is not the case the standard equation of a hyperbola above is defined for x real and y real.   What then, is the domain and the range?

The domain is: $x \le -a + u$ or $x \ge a + u$, and the range is: a set of all real numbers.

So more specifically, we can put the standard equation of a hyperbola the way below:

$$\dfrac{(x-u)^2}{a^2} - \dfrac{(y-v)^2}{b^2} = 1, \text{ where } a \text{ and } b > 0, \text{ and } x \le -a + u \text{ or } x \ge a + u.$$

And also, like an ellipse, a hyperbola has an <u>eccentricity</u>.
Unlike an ellipse however, the eccentricity is bigger than 1.   What is an eccentricity?

An eccentricity is <u>a ratio of a focal distance to a semi major axis</u>, so in the case of a hyperbola, the eccentricity is <u>the focal distance over half the transverse axis</u>.

So of a hyperbola, the eccentricity specifies the degree, to which the branches bend.
Thus, it can tell us how flat the branches are in a hyperbola or how sharp the vertices are.

An eccentricity is denoted by e, and we have: $e > 1$ for a hyperbola.

If the eccentricity e is close to 1, the foci are close to the vertices, and the branches bend a lot, so the vertices are very sharp.
Therefore, the bigger the eccentricity, the flatter the branches.
So if e is very large, the two branches look like two parallel lines.

And a hyperbola has a physical property as follows.

For instance, a light beam directed to one focus of a hyperbolic mirror is reflected toward the other focus. Such a property can be used for the construction of a reflecting telescope, and the idea of a hyperbola is used in a navigation system.

Fig. G

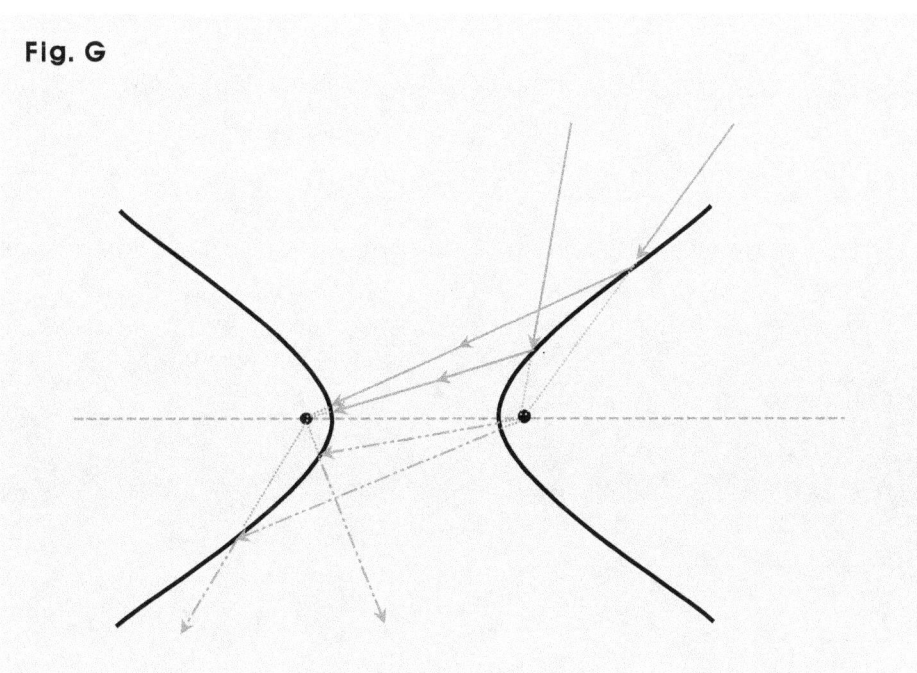

How then, can we get the foci and the eccentricity?

Suppose a hyperbola is: $\dfrac{(x-u)^2}{a^2} - \dfrac{(y-v)^2}{b^2} = 1$.

Then, the foci are: $(u - c, v)$ and $(u + c, v)$, where $c^2 = b^2 + a^2$, and c is the focal distance, that is, half the diagonal of the rectangle (the radius of the circumcircle).

Then, the eccentricity $e = c/a$.   And the hyperbola is horizontal.

Suppose however, a hyperbola is: $\dfrac{(y-v)^2}{b^2} - \dfrac{(x-u)^2}{a^2} = 1$.

Then, the foci are: $(u, v + c)$ and $(u, v - c)$, where $c^2 = b^2 + a^2$, and c is the focal distance.

Then, the eccentricity $e = c/b$.   And the hyperbola is vertical.

Examples 1 in Standard Forms

Label each hyperbola below, and then find the equation of each hyperbola.

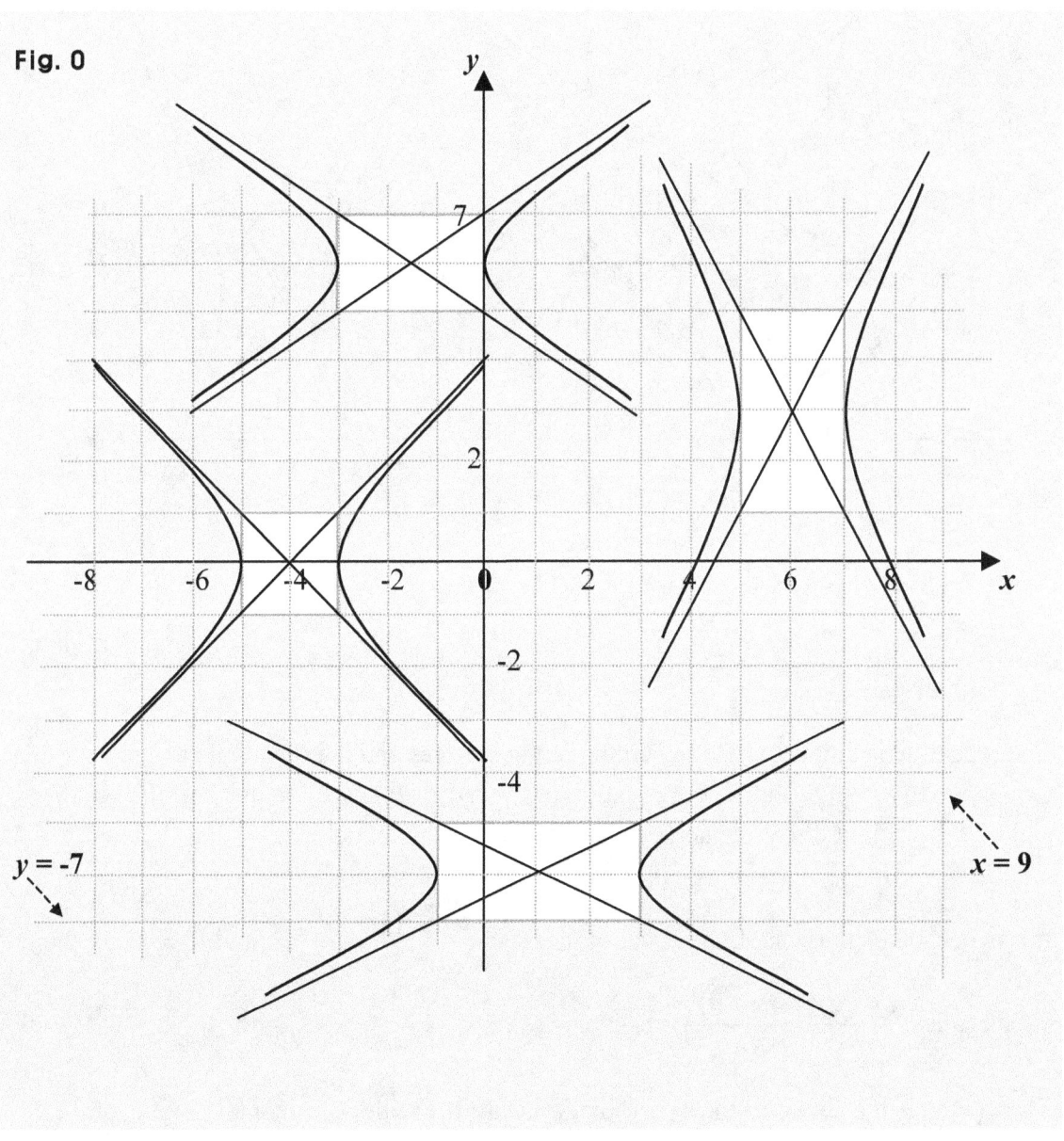

Fig. 0

Suggestions or Solutions
To the Problems in the Examples

The hyperbolas are all horizontal, and beginning with the hyperbola A below, we can see first, the center is $(-\frac{3}{2}, 6)$, and the rectangle K is 3 by 2.

Fig. 1

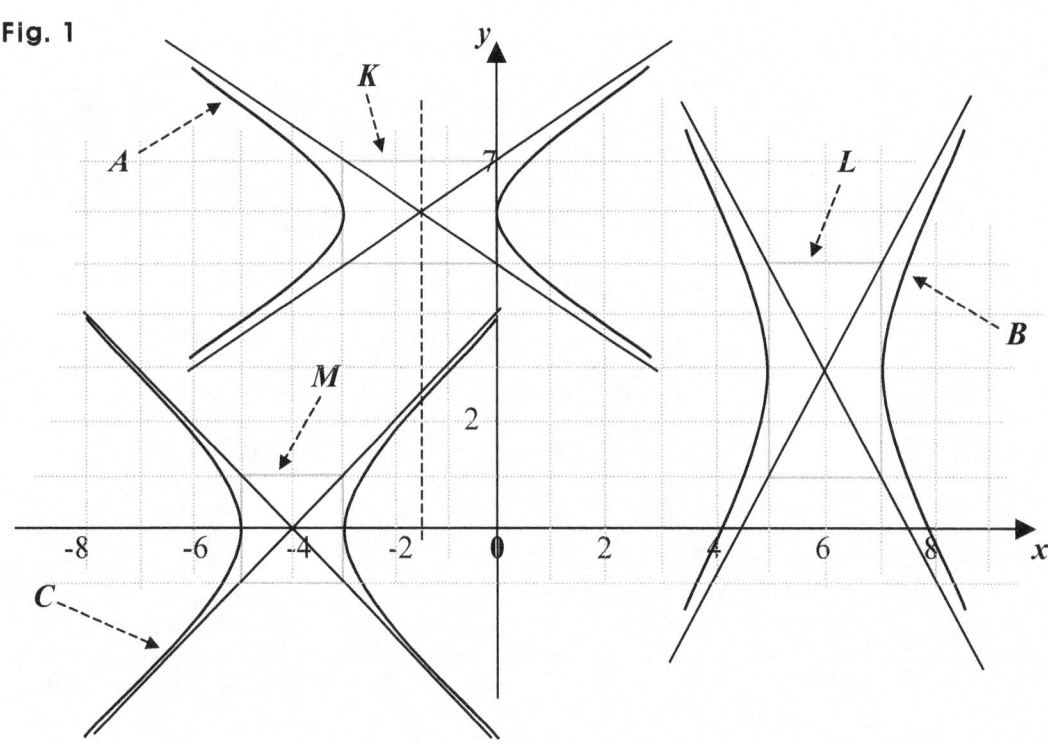

The rectangle is tangent to the hyperbola at the vertices and the diagonals overlap the two asymptotes. And finding the equation of the hyperbola, we can use <u>half the width and half the height</u> of the rectangle. It's because of the fact below:

If a hyperbola is horizontal, the equation is: $\dfrac{(x-u)^2}{a^2} - \dfrac{(y-v)^2}{b^2} = 1$, and if vertical, the

equation is: $\dfrac{(y-v)^2}{b^2} - \dfrac{(x-u)^2}{a^2} = 1$.

Then (u, v) is the center of the hyperbola, a is half the width (horizontal length) of the rectangle, and b is the half the height (vertical length).

And thus, beginning with the hyperbola A, we can see the center is $(-\frac{3}{2}, 6)$, and of the rectangle K, the width is 3, and the height is 2. So we get: $u = -\frac{3}{2}$, $v = 6$, $a = \frac{3}{2}$, and $b = 1$.

Thus, we get: $\dfrac{\{x-(-\frac{3}{2})\}^2}{(\frac{3}{2})^2} - \dfrac{(y-6)^2}{1^2} = 1 \Rightarrow \dfrac{(x+\frac{3}{2})^2}{(\frac{3}{2})^2} - \dfrac{(y-6)^2}{1^2} = 1$, which is the equation

of the hyperbola A. And we can put it this way, too, of course: $\dfrac{(x+\frac{3}{2})^2}{\frac{9}{4}} - (y-6)^2 = 1$.

And this way, also: $\dfrac{4(x+\frac{3}{2})^2}{9} - (y-6)^2 = 1$ or $4(x+\frac{3}{2})^2 - 9(y-6)^2 = 9$.

Next, moving on to the hyperbola B, we can see the center is $(6, 3)$, and of the rectangle L, the width is 2, and the height is 4. So we get: $u = 6$, $v = 3$, $a = 1$, and $b = 2$.

Thus, we get: $\dfrac{(x-6)^2}{1^2} - \dfrac{(y-3)^2}{2^2} = 1$, which is the equation of the hyperbola B.

And we can put it this way, too: $(x-6)^2 - \dfrac{(y-3)^2}{4} = 1$ or $4(x-6)^2 - (y-3)^2 = 4$.

Moving next, on to the hyperbola C, we can see the center is $(-4, 0)$, and of the rectangle M, the width is 2, and the height is 2. So we get: $u = -4$, $v = 0$, $a = 1$, and $b = 1$.

Thus, we get: $\dfrac{\{x-(-4)\}^2}{1^2} - \dfrac{(y-0)^2}{1^2} = 1 \Rightarrow \dfrac{(x+4)^2}{1^2} - \dfrac{y^2}{1^2} = 1$, which is the equation of the hyperbola C.

And of course, we can put it this way, too: $(x+4)^2 - y^2 = 4$.

And next, moving on to the other hyperbola, we have:

Fig. 2

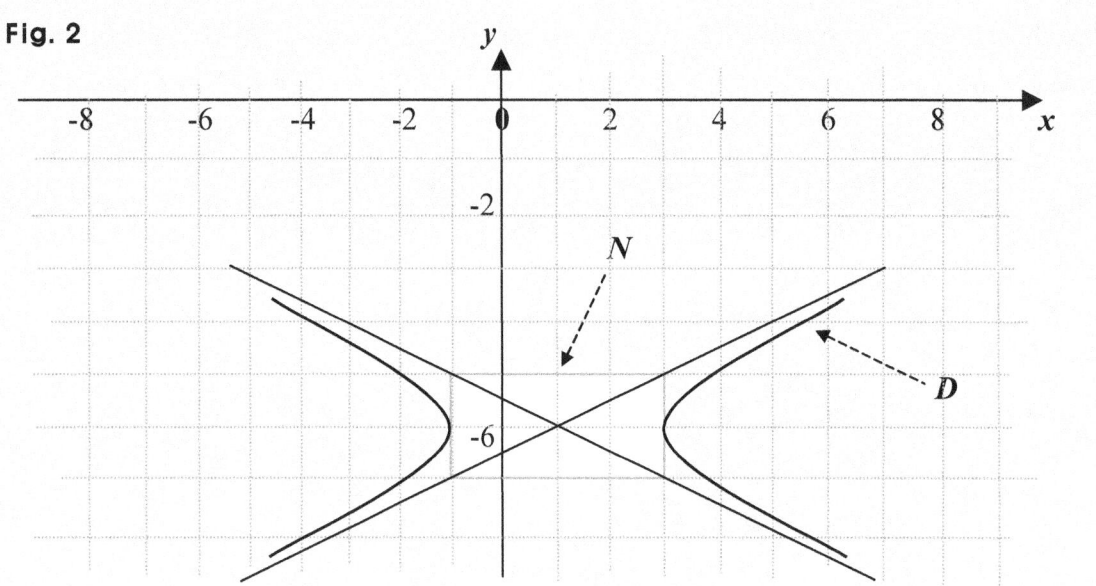

Then, we can see the center of hyperbola **D** is (1, -6), and of the rectangle **N**, the width is 4, and the height is 2. So we get: **u = 1**, **v = -6**, **a = 2**, and **b = 1**.

Thus, we get: $\dfrac{(x-1)^2}{2^2} - \dfrac{(y+6)^2}{1^2} = 1$, which is the equation of the hyperbola **D**.

And we can put it this way, too: $\dfrac{(x-1)^2}{4} - (y+6)^2 = 1$ or $(x-1)^2 - 4(y+6)^2 = 4$.

1. Equations for Hyperbolas 1

To begin with, putting a hyperbola in an equation, we can get: $\dfrac{(x-5)^2}{3^2} - \dfrac{(y-2)^2}{2^2} = 1.$

What hyperbola then, is it?

It is a hyperbola <u>horizontal</u>, the <u>center is at (5, 2)</u>, <u>the major axis is 6, and the minor is 4</u>. So the transverse axis is parallel to the *x*-axis, and is 6, and the hyperbola is as below:

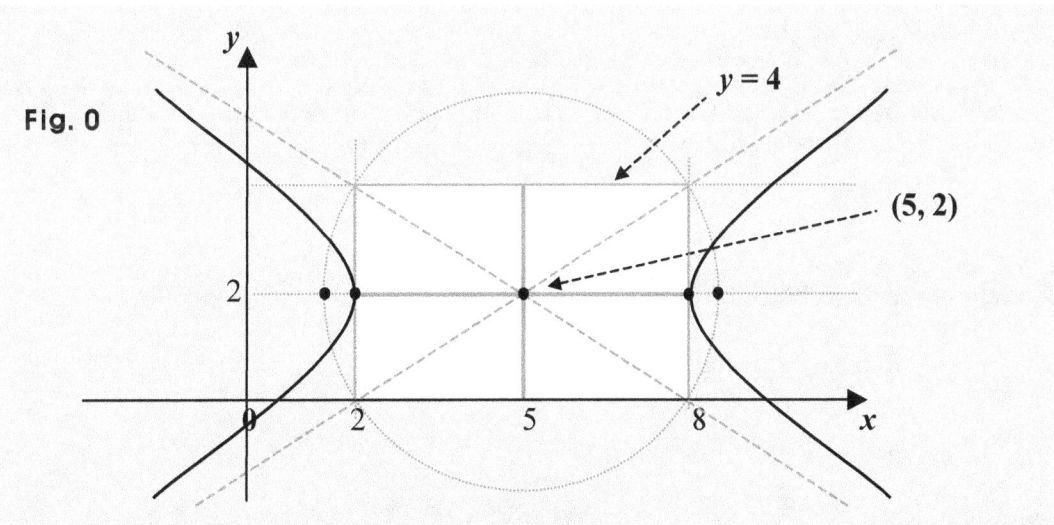

Fig. 0

The transverse axis has vertices as its endpoints. And the minor axis is called the conjugate axis, so the conjugate axis is 4, and is the line segment having the center as the midpoint and parallel to the *y*-axis.   What hyperbola then, is the hyperbola below?

$$\dfrac{x^2}{3^2} - \dfrac{y^2}{2^2} = 1.$$

The hyperbola is <u>horizontal</u>, too. And the <u>transverse axes is 6,</u> and the <u>conjugate is 4,</u> also, but the <u>center is different, and is at (0, 0)</u>, that is, at the origin.

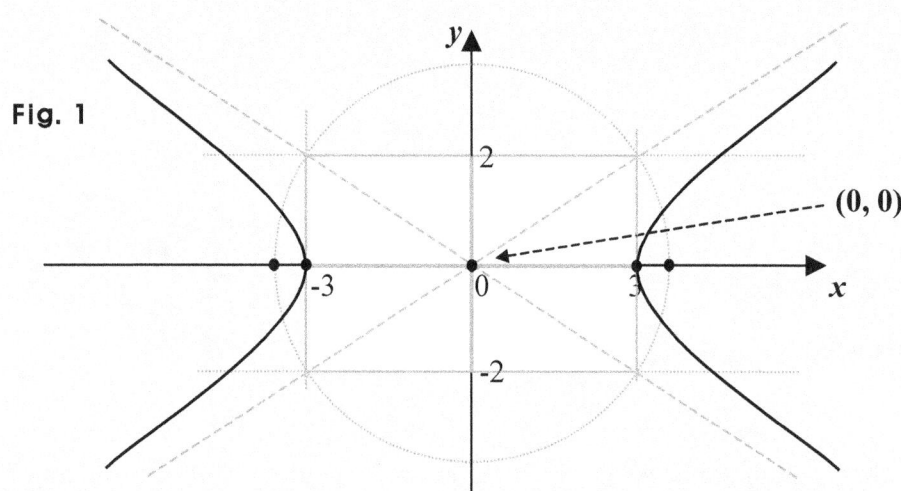

What if the center is at (1, 1), and the other information is the same?

Then, the hyperbola will be as follows: $\dfrac{(x-1)^2}{3^2} - \dfrac{(y-1)^2}{2^2} = 1$.

And we can put the hyperbola above in a graph the way below:

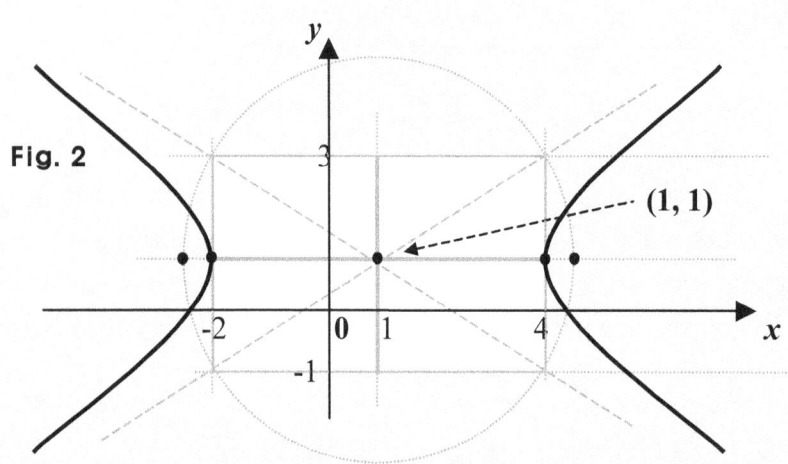

How then, can we get such equations as above?

We know the fact that a hyperbola is a set of points, from each of which, the difference between the two distances to two points called the foci is constant.   So?

So using the fact above, we can get the equation of a hyperbola.

Thus first, suppose $c > 0$, two points $P(-c, 0)$ and $Q(c, 0)$ are the two foci of a hyperbola called H, and a point $T(x, y)$ is an arbitrary point in the hyperbola H, that is, a random point representing all the points in H.

Then, the center is the origin, $(0, 0)$, and c is the focal distance, because the center is the midpoint between the foci. So we can put the three points P, Q, and T in the x-y plane the way as follows:

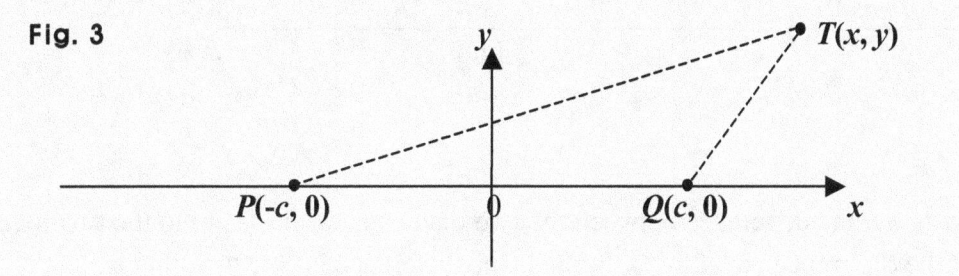

Suppose next, the point T is moving along the hyperbola H, and is now at $(a, 0)$. Then, we get:

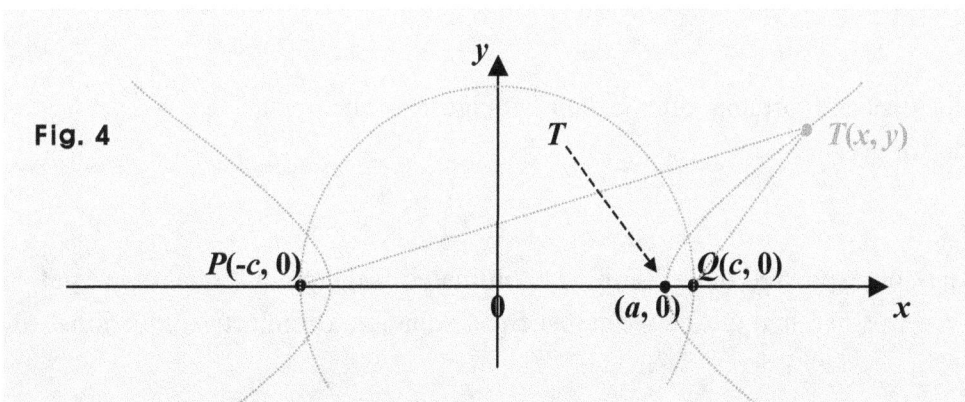

Then, the difference between the two distances from $T(x, y)$ to the two foci is as follows:

$\overline{TP} = a - (-c) = a + c$, and $\overline{TQ} = c - a$.   So $\overline{TP} - \overline{TQ} = a + c - (c - a) = 2a$.

And we know the fact that a hyperbola is a set of points, from each of which, the difference between the two distances to two points called the foci is constant.   So?

So no matter where the point *T* may be in the hyperbola *H*, the difference between the two distances from the point *T* to the two foci is always the same, and is in this case, **2a**.

Using thus, the fact above, we can get the equation of *H*. How then, can we use the fact?

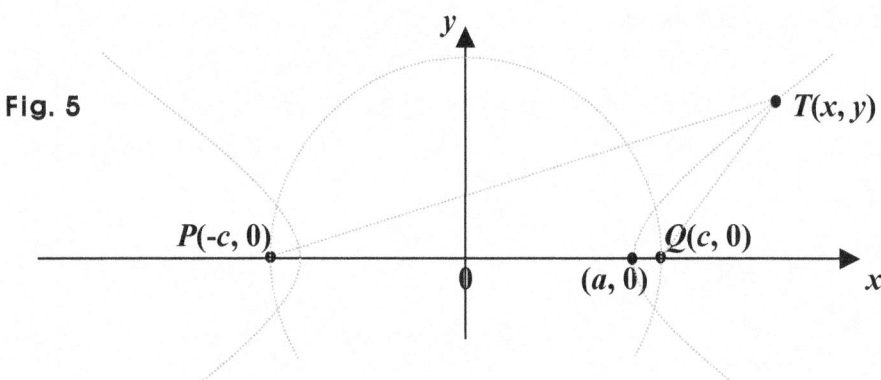

Fig. 5

We are going to get the difference between the two distances from *T(x, y)* to the two foci, and then, set the difference equal to **2a**, because the difference is constant, that is, the same no matter where the point *T* may be in the hyperbola *H*.

How then, can we get the two distances from *T(x, y)* to the two foci?

We can use the distance formula, often called Pythagorean Theorem.
What then, do we get?

We get an equation expressed in terms of the coordinates of the point *T*, that is, we get an equation in terms of *x* and *y*. And we call such an equation a connective equation.

So in this case, the connective equation is the equation that connects the two variables used as the coordinates of the arbitrary point *T(x, y)* in the curve called the hyperbola *H*.

How then, do we call the equation?

The equation explains every point in the hyperbola **H**, so it indicates the hyperbola **H**. The equation is thus, called the equation of the hyperbola **H**.

So let's now get the equation.   How then, can we apply the distance formula?

Assuming **d** is the distance between two points, **Δx** is the difference in **x**-coordinates, and **Δy** is the difference in **y**-coordinates, we can put the formula the way below:

$d^2 = (\Delta x)^2 + (\Delta y)^2.$

So using the distance formula, and assuming **TP** = **p**, and **TQ** = **q**, we can get:

$p^2 = \{x - (-c)\}^2 + (y - 0)^2 = (x + c)^2 + y^2$, and $q^2 = (x - c)^2 + (y - 0)^2 = (x - c)^2 + y^2.$

So we get: $TP = \sqrt{(x+c)^2 + y^2}$, and $TQ = \sqrt{(x-c)^2 + y^2}.$

Thus, we get: $TP - TQ = \sqrt{(x+c)^2 + y^2} - \sqrt{(x-c)^2 + y^2} = 2a.$ Then, we get:

$$\sqrt{(x+c)^2 + y^2} = 2a + \sqrt{(x-c)^2 + y^2} \Rightarrow (x+c)^2 + y^2 = (2a + \sqrt{(x-c)^2 + y^2})^2$$

$$\Rightarrow (x+c)^2 + y^2 = 4a^2 + 4a\sqrt{(x-c)^2 + y^2} + (x-c)^2 + y^2$$

$$\Rightarrow 4a\sqrt{(x-c)^2 + y^2} = (x+c)^2 + y^2 - (x-c)^2 - y^2 - 4a^2$$

$$\Rightarrow 4a\sqrt{(x-c)^2 + y^2} = x^2 + 2cx + c^2 - x^2 + 2cx - c^2 - 4a^2 = 4cx - 4a^2$$

$$\Rightarrow a\sqrt{(x-c)^2 + y^2} = cx - a^2 \Rightarrow a^2\{(x-c)^2 + y^2\} = (cx - a^2)^2 = c^2x^2 - 2a^2cx + a^4$$

$$\Rightarrow a^2(x^2 - 2cx + c^2 + y^2) = a^2x^2 - 2a^2cx + a^2c^2 + a^2y^2 = c^2x^2 - 2a^2cx + a^4$$

$$\Rightarrow a^2x^2 - c^2x^2 + a^2c^2 + a^2y^2 = a^4 \Rightarrow (a^2 - c^2)x^2 + a^2y^2 = a^4 - a^2c^2 = a^2(a^2 - c^2)$$

$$\Rightarrow x^2 + \frac{a^2y^2}{a^2 - c^2} = a^2 \Rightarrow \frac{x^2}{a^2} + \frac{y^2}{a^2 - c^2} = 1.$$ What then, about $a^2 - c^2$?

We can add to the graph a circle of radius *c* centered at the origin the way below:

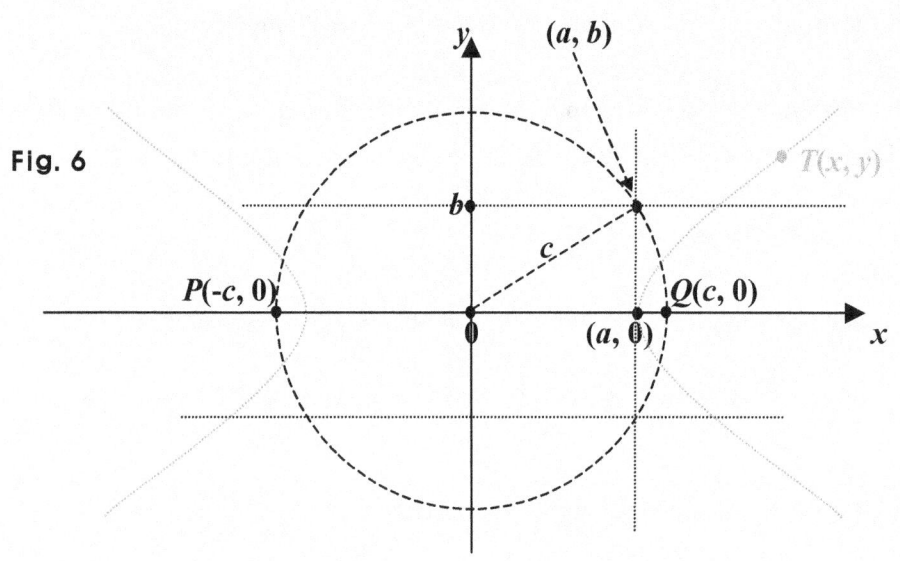

Fig. 6

Then, the circle's center is the center of the hyperbola.   So?

So the radius is the focal distance of the hyperbola.   What then?

Assuming **b > 0**, and putting a point **(a, b)** in the circle the way above, we can get a right triangle where the vertices are **(a, 0)**, **(a, b)**, and the origin. So the three sides in the right triangle are **a**, **b**, and **c**, where **c** is the hypotenuse.   So?

Applying thus, the distance formula, we can get this: $c^2 = b^2 + a^2 \Rightarrow b^2 = c^2 - a^2$.

And we now have: $\dfrac{x^2}{a^2} + \dfrac{y^2}{a^2 - c^2} = 1$.   So we get: $\dfrac{x^2}{a^2} - \dfrac{y^2}{b^2} = 1$.

And the equation above is an equation expressed in terms of *x* and *y*. And we call such an equation a connective equation connecting *x* and *y*.

So in this case, the connective equation is the equation that connects the two variables used as the coordinates of the arbitrary point *T(x, y)* in the curve called the hyperbola *H*.

Thus, the equation of the hyperbola H is: $\dfrac{x^2}{a^2} - \dfrac{y^2}{b^2} = 1$.

And putting the hyperbola H in the x-y plane, we get:

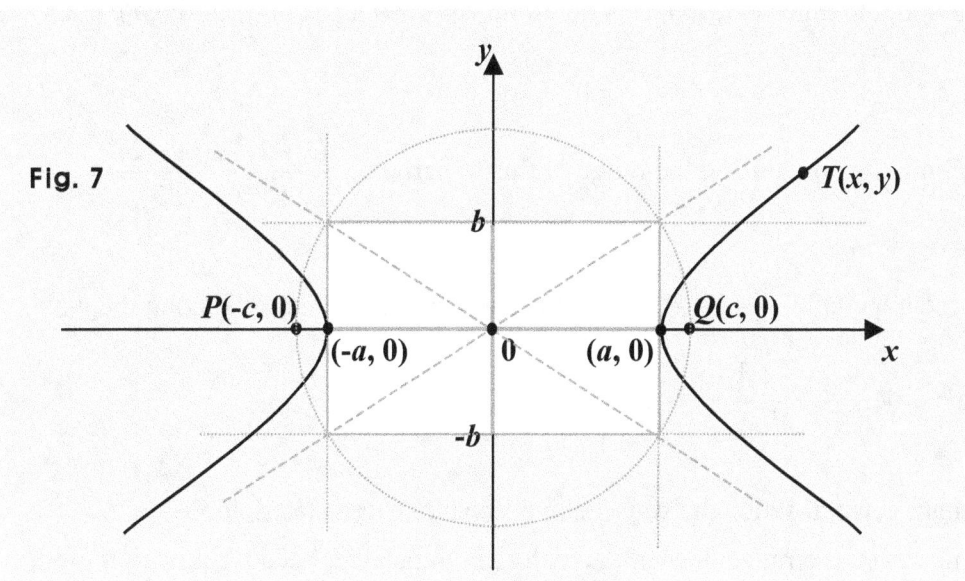

Fig. 7

So we can now say that the hyperbola H is horizontal and centered at the origin, the major axis is $2a$, and is parallel to the x-axis, and the minor axis is $2b$.

That is, the transverse axis is $2a$, and is horizontal, and the conjugate axis $2b$.

Then, putting the hyperbola H in the equation, we get: $\dfrac{x^2}{a^2} - \dfrac{y^2}{b^2} = 1$.

What then, are the domain and the range?

The domain is: $x \le -a$ or $x \ge a$, that is, $|x| \ge a$. And the range is: $y \in R$ where R is a set of all real numbers, that is, the range is a set of all real numbers.

What then, about the domain and the range of this hyperbola: $\dfrac{x^2}{2^2} - \dfrac{y^2}{3^2} = 1$?

The domain is $x \leq -2$ or $x \geq 2$, that is, $|x| \geq 2$.

And the range is: $y \in R$ where R is a set of all real numbers, that is, the range is a set of all real numbers.

This time though, the transverse axis is 4, and thus, is smaller than the conjugate axis, which is 6.

What then, about the domain and the range of this hyperbola: $\dfrac{(x-1)^2}{2^2} - \dfrac{(y-2)^2}{3^2} = 1$?

Translating the hyperbola $\dfrac{x^2}{2^2} - \dfrac{y^2}{3^2} = 1$ by 1 along the x-axis, and by 2 along the y-axis, we get the hyperbola $\dfrac{(x-1)^2}{2^2} - \dfrac{(y-2)^2}{3^2} = 1$.

And the domain gets translated the way the hyperbola gets translated, too.
In this case however, the range does not actually get translated, because it is a set of all real numbers. So the range remains the same. And getting the new domain, we get:

$x \leq -2 + 1 = -1$ or $x \geq 2 + 1 = 3$, so the new domain is: $x \leq -1$ or $x \geq 3$.

What then, about the domain and the range of this hyperbola: $\dfrac{(x+1)^2}{2^2} - \dfrac{(y+2)^2}{3^2} = 1$?

Translating this time, the hyperbola $\dfrac{x^2}{2^2} - \dfrac{y^2}{3^2} = 1$ by -1 along the x-axis, and by -2 along the y-axis, we get this hyperbola $\dfrac{(x+1)^2}{2^2} - \dfrac{(y+2)^2}{3^2} = 1$.

And the domain gets translated the way the hyperbola gets translated, too.
In this case, too, though, the range does not actually get translated, because it is a set of all real numbers. So the range remains the same. And getting the new domain, we get:

$x \leq -2 - 1 = -3$ or $x \geq 2 - 1 = 1$, so the new domain is: $x \leq -3$ or $x \geq 1$.

What then, about this hyperbola: $\dfrac{(x-u)^2}{a^2} - \dfrac{(y-v)^2}{b^2} = 1$?

Translating the hyperbola $\dfrac{x^2}{a^2} - \dfrac{y^2}{b^2} = 1$ by u along the x-axis, and by v along the y-axis,

we get this hyperbola: $\dfrac{(x-u)^2}{a^2} - \dfrac{(y-v)^2}{b^2} = 1$.

And the domain gets translated the way the hyperbola gets translated, too.

We know that the domain of $\dfrac{x^2}{a^2} - \dfrac{y^2}{b^2} = 1$ is: $x \le \text{-}a$ or $x \ge a$.

So getting the domain of $\dfrac{(x-u)^2}{a^2} - \dfrac{(y-v)^2}{b^2} = 1$, we get: $x \le \text{-}a + u$, or $x \ge a + u$.

In this case, too, though, the range does not actually get translated, because it is a set of all real numbers. So the range remains the same.

And note that unlike an ellipse, the major axis of a hyperbola, that is, the transverse axis does not have to be bigger than the minor axis, which is often called the conjugate axis.

So the transverse axis does not have to be larger than the conjugate axis.
In other words, the conjugate axis can be larger than the transverse axis, too.

Examples 2 in Standard Forms

Label each hyperbola below, and then find the equation of each hyperbola.

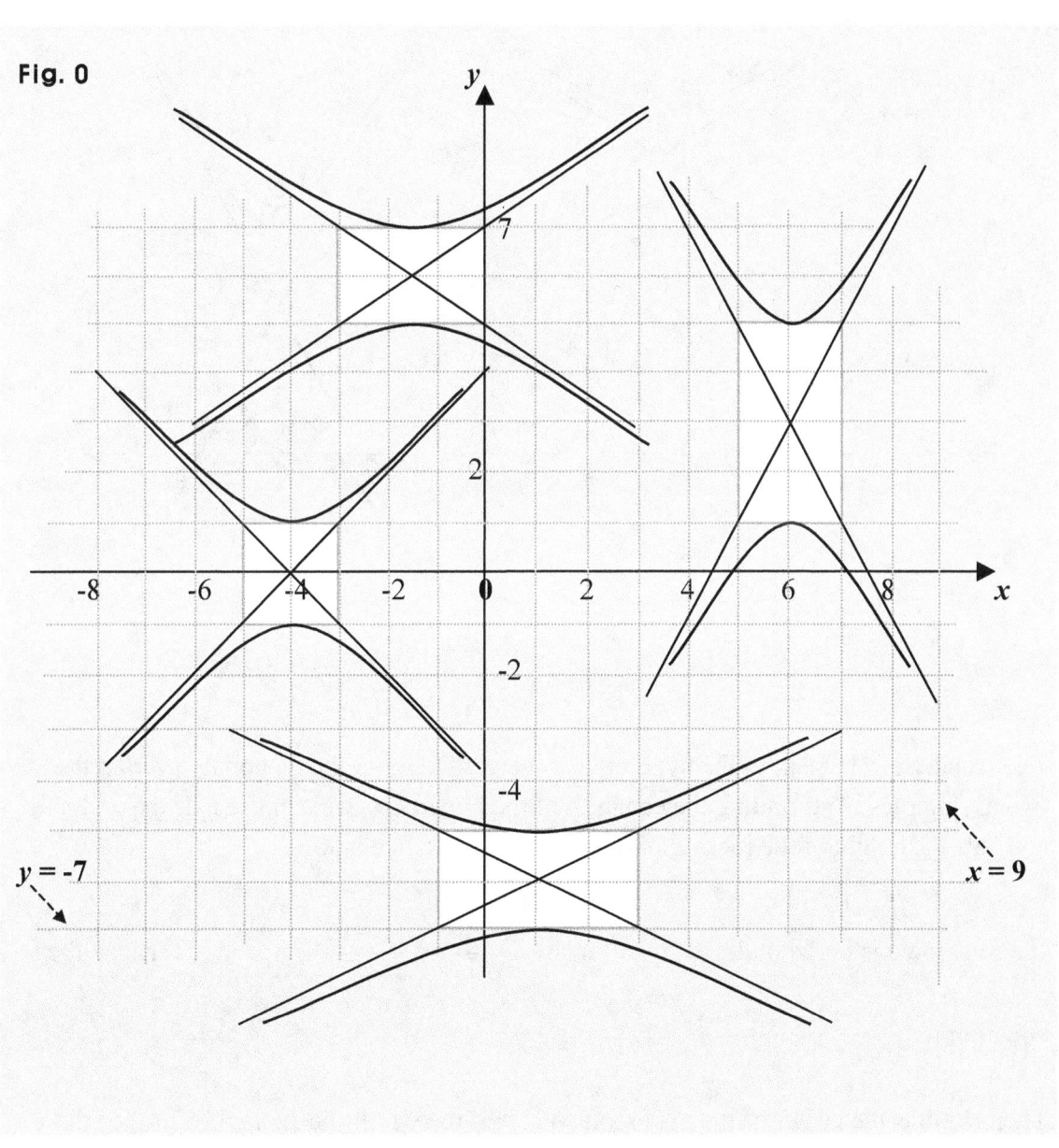

Suggestions or Solutions
To the Problems in the Examples

The hyperbolas are all vertical, and beginning with the hyperbola *A* below, we can see first, the center is $(-\frac{3}{2}, 6)$, and the rectangle *K* is 3 by 2.

Fig. 1

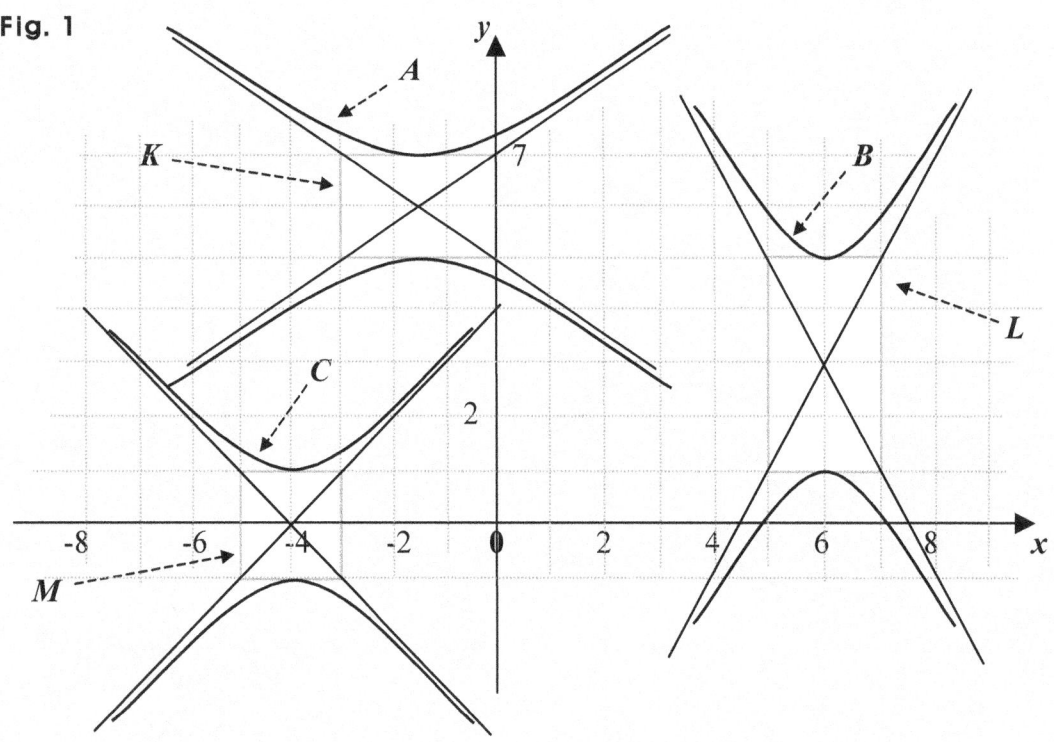

The rectangle is tangent to the hyperbola at the vertices, and the diagonals overlap the two asymptotes. And finding the equation of the hyperbola, we can use <u>half the width and half the height</u> of the rectangle. It's because of the fact below:

If a hyperbola is horizontal, the equation is: $\dfrac{(x-u)^2}{a^2} - \dfrac{(y-v)^2}{b^2} = 1$, and if vertical, the equation is: $\dfrac{(y-v)^2}{b^2} - \dfrac{(x-u)^2}{a^2} = 1$.

Then **(u, v)** is the center of the hyperbola, **a** is half the width (horizontal length) of the rectangle, and **b** is the half the height (vertical length).

And thus, beginning with the hyperbola A, we can see the center is $(-\frac{3}{2}, 6)$, and of the rectangle K, the width is 3, and the height is 2. So we get: $u = -\frac{3}{2}$, $v = 6$, $a = \frac{3}{2}$, and $b = 1$.

Thus, we get: $\dfrac{(y-6)^2}{1^2} - \dfrac{\{x-(-\frac{3}{2})\}^2}{(\frac{3}{2})^2} = 1 \Rightarrow \dfrac{(y-6)^2}{1^2} - \dfrac{(x+\frac{3}{2})^2}{(\frac{3}{2})^2} = 1$, which is the equation

of the hyperbola A. And we can put it this way, too, of course: $(y-6)^2 - \dfrac{(x+\frac{3}{2})^2}{\frac{9}{4}} = 1$.

And this way, also: $(y-6)^2 - \dfrac{4(x+\frac{3}{2})^2}{9} = 1$ or $9(y-6)^2 - 4(x+\frac{3}{2})^2 = 9$.

Next, moving on to the hyperbola B, we can see the center is $(6, 3)$, and of the rectangle L, the width is 2, and the height is 4. So we get: $u = 6$, $v = 3$, $a = 1$, and $b = 2$.

Thus, we get: $\dfrac{(y-3)^2}{2^2} - \dfrac{(x-6)^2}{1^2} = 1$, which is the equation of the hyperbola B.

And we can put it this way, too: $\dfrac{(y-3)^2}{4} - (x-6)^2 = 1$ or $(y-3)^2 - 4(x-6)^2 = 4$.

Moving next, on to the hyperbola C, we can see the center is $(-4, 0)$, and of the rectangle M, the width is 2, and the height is 2. So we get: $u = -4$, $v = 0$, $a = 1$, and $b = 1$.

Thus, we get: $\dfrac{(y-0)^2}{1^2} - \dfrac{\{x-(-4)\}^2}{1^2} = 1 \Rightarrow \dfrac{y^2}{1^2} - \dfrac{(x+4)^2}{1^2} = 1$, which is the equation of the hyperbola C.

And of course, we can put it this way, too: $y^2 - (x+4)^2 = 4$.

And next, moving on to the other hyperbola, we have:

Fig. 2

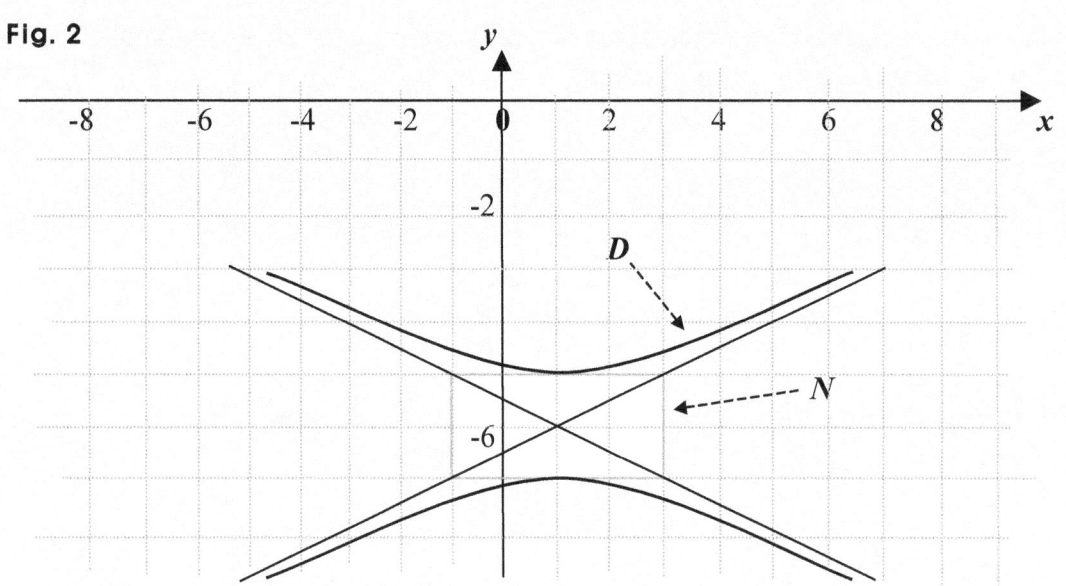

Then, we can see the center of hyperbola **D** is (1, -6), and of the rectangle **N**, the width is 4, and the height is 2. So we get: $u = 1$, $v = -6$, $a = 2$, and $b = 1$.

Thus, we get: $\dfrac{(y+6)^2}{1^2} - \dfrac{(x-1)^2}{2^2} = 1$, which is the equation of the hyperbola **D**.

And we can put it this way, too: $(y+6)^2 - \dfrac{(x-1)^2}{4} = 1$ or $4(y+6)^2 - (x-1)^2 = 4$.

₂.Equations for Hyperbolas 2

To begin with, putting a hyperbola in an equation, we can get: $\dfrac{x^2}{a^2} - \dfrac{y^2}{b^2} = 1$.

What hyperbola then, is it?

It is a hyperbola <u>horizontal</u>, where the <u>center is at (0, 0)</u>, and its <u>transverse axis is **2a**</u> and is parallel to the **x**-axis, and <u>the conjugate axis is **2b**</u>.   And the hyperbola is as follows:

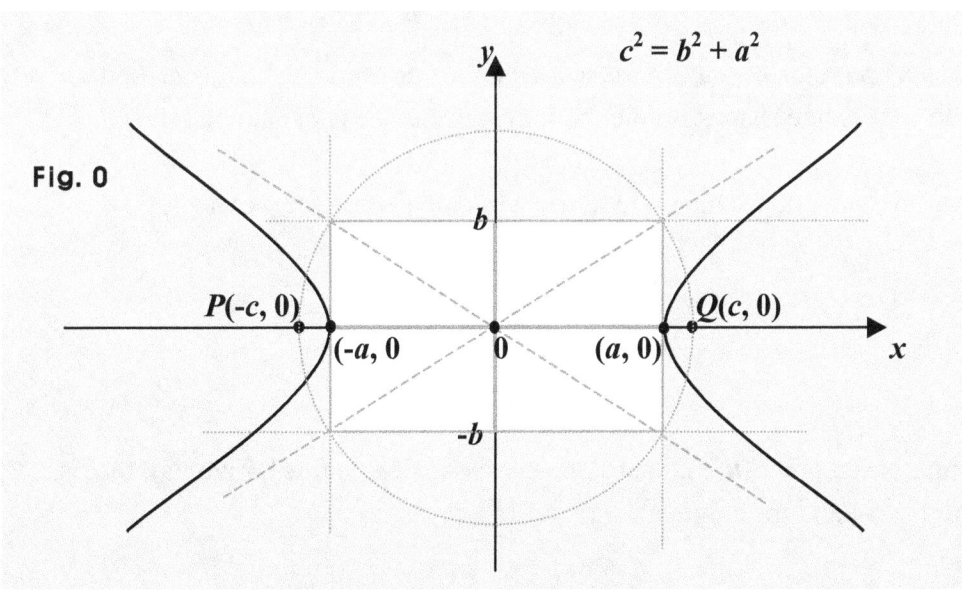

Fig. 0

What if the hyperbola above is centered at a point other than the origin?
For instance, moving the center to (5, 1), we get a new hyperbola, so we get a new equation. What then, will the new equation be?

36

Assuming the hyperbola above is **H**, and the new hyperbola is **G**, we can get **G** translating **H** in the amount of 5 along the *x*-axis, and in the amount of 2 along the *y*-axis.

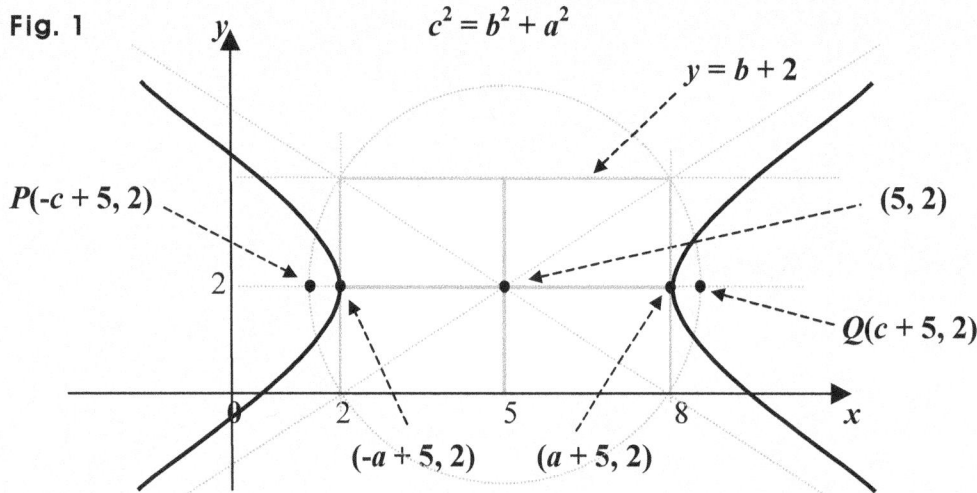

Fig. 1

$c^2 = b^2 + a^2$

$y = b + 2$

$P(-c + 5, 2)$

$(5, 2)$

$Q(c + 5, 2)$

$(-a + 5, 2)$ $(a + 5, 2)$

Then, the equation of the hyperbola **G** is: $\dfrac{(x-5)^2}{a^2} - \dfrac{(y-2)^2}{b^2} = 1$.

(Note that we still get: $c^2 = b^2 + a^2$, because **a** and **b** are constant, and so is **c**. And translations do not change shapes, so the focal distance **c** does not change either.)

And in general, we can put a standard hyperbola in an equation the way as follows:

$$\frac{(x-u)^2}{a^2} - \frac{(y-v)^2}{b^2} = 1.$$

Then, the center is at a point **(u, v)**, the transverse axis is **2a**, and is horizontal, that is, parallel to the *x*-axis, and the conjugate axis is: **2b**.

And the equation above is called the standard equation of a hyperbola.
(Note however, that the standard equation above can indicate a hyperbola standard only. If a hyperbola is standard, the transverse axis is parallel to a coordinate axis.)

What if a hyperbola is not standard, thus, tilted or slanted as the hyperbola below?

Fig. 2

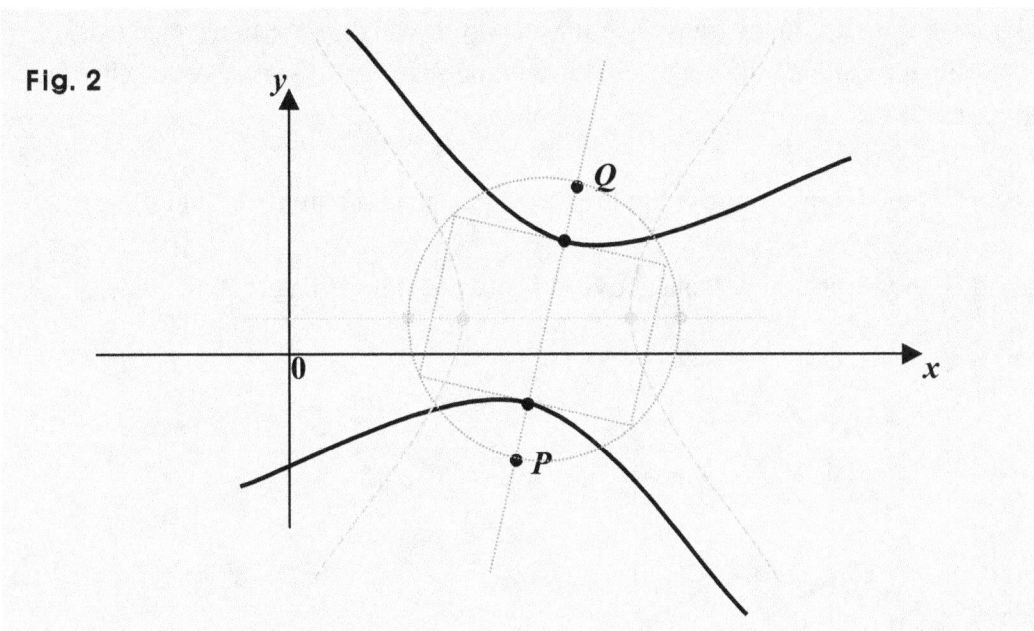

Given the foci, you can derive the equation using the definition for hyperbolas, or can get the equation applying a transformation to a hyperbola standard. The transformation is covered in the section, Transformation by Turning, in the book, **GRAPH OPERATIONS**.

What then, about the ellipse described below?

The <u>center is at (0, 0)</u>, the <u>transverse axis</u> is this time, **2b**, and is <u>parallel to the y-axis</u>, and <u>the conjugate axis</u> is **2a**.

Then, the hyperbola is <u>vertical</u>, and the equation is: $\dfrac{y^2}{b^2} - \dfrac{x^2}{a^2} = 1$.

So this time, the signs of the quadratic terms got swapped, and the equation looks quite opposite of the equation of a hyperbola horizontal.

What then, are the domain and the range?

A domain is a set of all the *x*-values, and a range is a set of all the *y*-values.

So the domain is a set of all real numbers, and the range is: $y \geq b$ or $y \leq -b$, i.e., $|y| \geq b$.

And we know the foci are in the transverse line, a part of which is the transverse axis, which is this time, parallel to the *y*-axis, and is the line segment from one vertex **(0, b)** to the other vertex **(0, -b)**.

And the center is the origin, so if *c* is the focal distance, the foci are **(0, c)** and **(0, -c)**.

Assuming thus, the vertical hyperbola above is *V*, we can put *V* in a graph the way below:

Fig. 3

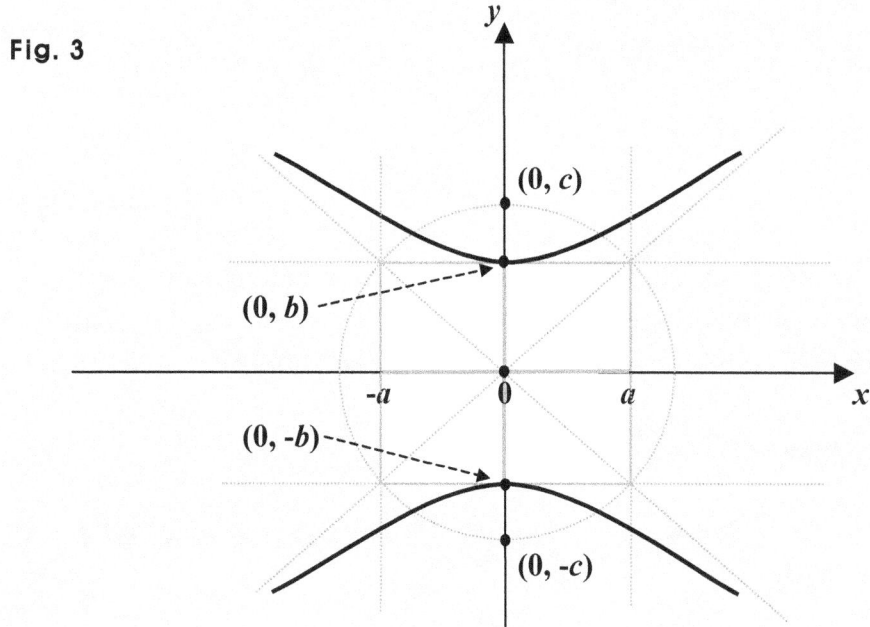

How then, can we get the equation of *V*?

We know the definition for hyperbolas is saying that a hyperbola is a set of points, from each of which, the difference between the two distances to the two foci is constant.

Using thus, the fact above again, we can get the equation of a hyperbola vertical, too.

So to begin with, suppose *c* > **0**, two points **P(0, c)** and **Q(0, -c)** are the two foci of a hyperbola called *V*, and a point **T(x, y)** is an arbitrary point in the hyperbola *V*, that is, a random point representing all the points in *V*.

Then, the center is the origin, that is, **(0, 0)**, and we can say that *c* is the focal distance, because the center is the midpoint between the two foci. So we can put the three points *P*, *Q*, and *T* in the *x-y* plane the way as follows:

Fig. 4

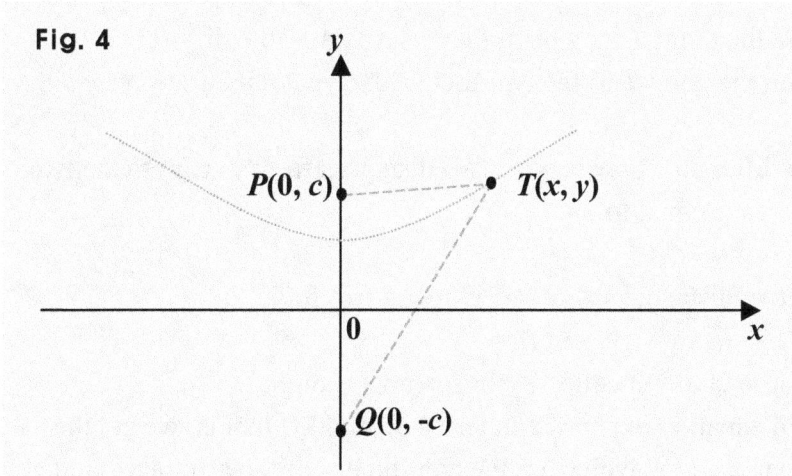

Suppose next, the point *T* is moving now, along the hyperbola *V*.

And we know that the difference between two distances from each point in *V* to the two foci is constant, that is, the same.   So this time, too, we are going to use the fact again.

So suppose this time, that the point *T* is now at the point **(0, b)**.

Fig. 5

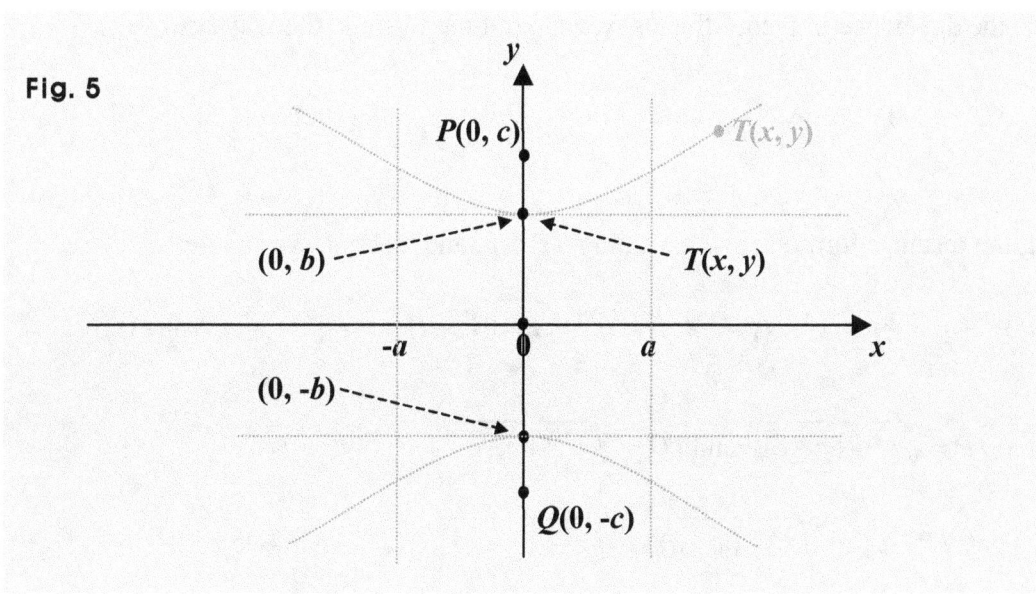

We can see that the length of **TQ** is: $b - (-c) = b + c$.

And the length of **TP** is: $c - b$.　So we get: $TQ - TP = (b + c) - (c - b) = 2b$.

Therefore, no matter where the point **T** may be in the hyperbola **V**, the difference between two distances from the point **T** to the two foci is always **2b**, the transverse axis.

So we are going to get the difference between the two distances from **T(x, y)** to the two foci, and then, set the difference equal to **2b**.

How then, can we get the two distances from **T(x, y)** to the two foci?

We can use the distance formula, often called Pythagorean Theorem.
And using it, we can get an equation expressed in terms of **x** and **y**, that is, we get the connective equation between **x** and **y**, which are the coordinates of the arbitrary point **T(x, y)** in the curve called the hyperbola **V**.　How then, do we call the equation?

The equation explains every point in the hyperbola **V**, so it indicates the hyperbola **V**.

The equation is thus, called the equation of the hyperbola **V**.
So let's now get the equation.　How then, can we apply the distance formula?

Assuming **d** is the distance between two points, **Δx** is the difference in **x**-coordinates, and **Δy** is the difference in **y**-coordinates, we can put the formula the way below:

$d^2 = (\Delta x)^2 + (\Delta y)^2$.　　And we have: **P(0, c)**, **Q(0, -c)**, and **T(x, y)**.

So using the distance formula, and assuming **TP = p**, and **TQ = q**, we can get:

$p^2 = (x - 0)^2 + (y - c)^2 = x^2 + (y - c)^2$, and $q^2 = (x - 0)^2 + \{y - (-c)\}^2 = x^2 + (y + c)^2$.

So we get: $TP = \sqrt{x^2 + (y - c)^2}$, and $TQ = \sqrt{x^2 + (y + c)^2}$.

Thus, we get: $TP - TQ = \sqrt{x^2 + (y - c)^2} - \sqrt{x^2 + (y + c)^2} = 2b$.　Then, we get:

$$\sqrt{x^2+(y-c)^2} = 2b + \sqrt{x^2+(y+c)^2} \Rightarrow x^2+(y-c)^2 = (2b+\sqrt{x^2+(y+c)^2})^2$$

$$\Rightarrow x^2+(y-c)^2 = 4b^2 + 4b\sqrt{x^2+(y+c)^2} + x^2+(y+c)^2$$

$$\Rightarrow 4b\sqrt{x^2+(y+c)^2} = (y-c)^2 - (y+c)^2 - 4b^2 = y^2 - 2cy + c^2 - y^2 - 2cy - c^2 - 4b^2$$

$$\Rightarrow 4b\sqrt{x^2+(y+c)^2} = -4b^2 - 4cy \Rightarrow b\sqrt{x^2+(y+c)^2} = -(b^2+cy)$$

$$\Rightarrow b^2\{x^2+(y+c)^2\} = (b^2+cy)^2 = b^4 + 2b^2cy + c^2y^2$$

$$\Rightarrow b^2(x^2+y^2+2cy+c^2) = b^2x^2 + b^2y^2 + 2b^2cy + b^2c^2 = a^4 + 2b^2cy + c^2y^2$$

$$\Rightarrow b^2x^2 + b^2y^2 + b^2c^2 - c^2y^2 = b^4 \Rightarrow b^2x^2 + (b^2-c^2)y^2 = b^4 - b^2c^2 = b^2(b^2-c^2)$$

$$\Rightarrow \frac{b^2x^2}{b^2-c^2} + y^2 = b^2 \Rightarrow \frac{x^2}{b^2-c^2} + \frac{y^2}{b^2} = 1. \quad \text{What then, about } b^2-c^2?$$

We can add to the graph a circle of radius c centered at the origin the way below:

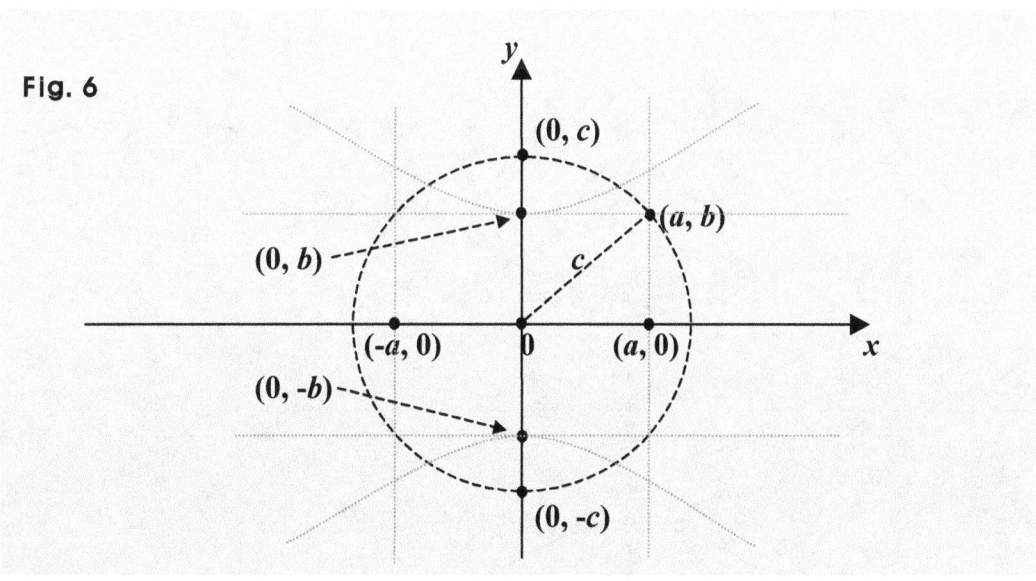

Fig. 6

Then, the circle's center is the center of the hyperbola.   So?

So the radius is the focal distance of the hyperbola.   What then?

Assuming $b > 0$, and putting a point (a, b) in the circle the way above, we can get a right triangle where the vertices are $(a, 0)$, (a, b), and the origin. So the three sides in the right triangle are a, b, and c, where c is the hypotenuse.   So?

Applying thus, the distance formula, we can get this: $c^2 = b^2 + a^2 \Rightarrow -a^2 = b^2 - c^2$.

And we now have: $\dfrac{x^2}{b^2 - c^2} + \dfrac{y^2}{b^2} = 1$.   So we get: $-\dfrac{x^2}{a^2} + \dfrac{y^2}{b^2} = 1 \Rightarrow \dfrac{y^2}{b^2} - \dfrac{x^2}{a^2} = 1$.

And the equation above is an equation expressed in terms of x and y. And we call such an equation a connective equation connecting x and y.

So in this case, the connective equation is the equation that connects the two variables used as the coordinates of the arbitrary point $T(x, y)$ in the curve called the hyperbola V.

Thus, the equation of the hyperbola V is: $\dfrac{y^2}{b^2} - \dfrac{x^2}{a^2} = 1$.

And putting the hyperbola V in the x-y plane, we get:

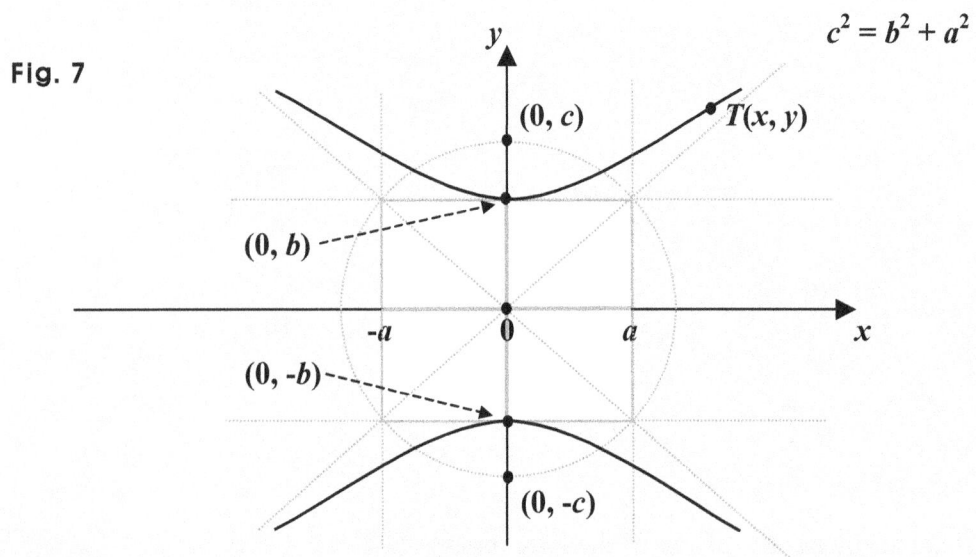

Fig. 7

So we can now say that the hyperbola V is vertical and centered at the origin, the major axis is **2b**, and is parallel to the y-axis, and the minor axis is **2a**.

That is, the transverse axis is **2b**, and is vertical, and the conjugate axis **2a**.

Then, putting the hyperbola V in the equation, we get: $\dfrac{y^2}{b^2} - \dfrac{x^2}{a^2} = 1$.

What are then, the domain and the range?

The domain is: $x \in R$ where R is a set of all real numbers, that is, the domain is a set of all real numbers. And the range is: $y \le -b$ or $y \ge b$, that is, $|y| \ge b$.

What then, about the domain and the range of this hyperbola: $\dfrac{y^2}{2^2} - \dfrac{x^2}{3^2} = 1$?

The domain is: $x \in R$ where R is a set of all real numbers, that is, the domain is a set of all real numbers. And the range is: $y \le -2$ or $y \ge 2$, that is, $|y| \ge 2$.

Note that the transverse axis is 4, and thus, is smaller than the conjugate axis, which is 6. So the transverse axis can be smaller than the conjugate axis.

What then, about the domain and the range of this hyperbola: $\dfrac{(y-1)^2}{2^2} - \dfrac{(x-2)^2}{3^2} = 1$?

Translating the hyperbola $\dfrac{y^2}{2^2} - \dfrac{x^2}{3^2} = 1$ by 2 along the x-axis, and by 1 along the y-axis,

we get the hyperbola $\dfrac{(y-1)^2}{2^2} - \dfrac{(x-2)^2}{3^2} = 1$.

And the range gets translated the way the hyperbola gets translated, too. And getting the new range, we get:

$y \le -2 + 1 = -1$ or $y \ge 2 + 1 = 3$, so the new range is: $y \le -1$ or $y \ge 3$.

In this case however, the domain does not actually get translated, because it is a set of all real numbers. So the domain remains the same, and thus, is a set of all real numbers.

What then, about the domain and the range of this hyperbola: $\dfrac{(y+1)^2}{2^2} - \dfrac{(x+2)^2}{3^2} = 1$?

Translating this time, the hyperbola $\dfrac{y^2}{2^2} - \dfrac{x^2}{3^2} = 1$ by -2 along the x-axis, and by -1 along the y-axis, we get this hyperbola: $\dfrac{(y+1)^2}{2^2} - \dfrac{(x+2)^2}{3^2} = 1$.

And the range gets translated the way the hyperbola gets translated, too. And getting the new range, we get: $y \leq -2 - 1 = -3$ or $y \geq 2 - 1 = 1$, so the new range is: $y \leq -3$ or $y \geq 1$.

In this case, too, though, the domain does not actually get translated, because it is a set of all real numbers. So the domain remains the same, thus, a set of all real numbers.

What then, about this hyperbola: $\dfrac{(y-v)^2}{b^2} - \dfrac{(x-u)^2}{a^2} = 1$?

Translating the hyperbola $\dfrac{y^2}{b^2} - \dfrac{x^2}{a^2} = 1$ by u along the x-axis, and by v along the y-axis, we get this hyperbola: $\dfrac{(y-v)^2}{b^2} - \dfrac{(x-u)^2}{a^2} = 1$.

And the range gets translated the way the hyperbola gets translated along the y-axis, too.

We know that the range of $\dfrac{x^2}{a^2} - \dfrac{y^2}{b^2} = 1$ is: $y \leq -b$ or $y \geq b$.

So getting the range of $\dfrac{(y-v)^2}{b^2} - \dfrac{(x-u)^2}{a^2} = 1$, we get: $y \leq -b + v$, or $y \geq b + v$.

In this case, too, though, the domain does not actually get translated, because it is a set of all real numbers. So the range remains the same, and is a set of all real numbers.

(And note that unlike an ellipse, the major axis of a hyperbola, that is, the transverse axis does not have to be bigger than the minor axis, which is often called the conjugate axis. So the transverse axis does not have to be larger than the conjugate axis. That is, it can be the case, too, the conjugate axis can be larger than the transverse axis.)

So in sum, we have just covered two equations indicating hyperbolas in two kinds.

One is: $\dfrac{x^2}{a^2} - \dfrac{y^2}{b^2} = 1$, where $c^2 = b^2 + a^2$, where c is the focal distance.

The hyperbola above is <u>horizontal</u>, the <u>center is at (0, 0)</u>, the <u>transverse axis is **2a**</u>, and is a part of (thus, parallel to) the **x**-axis, and the conjugate axis is **2b**. So the hyperbola is as below:

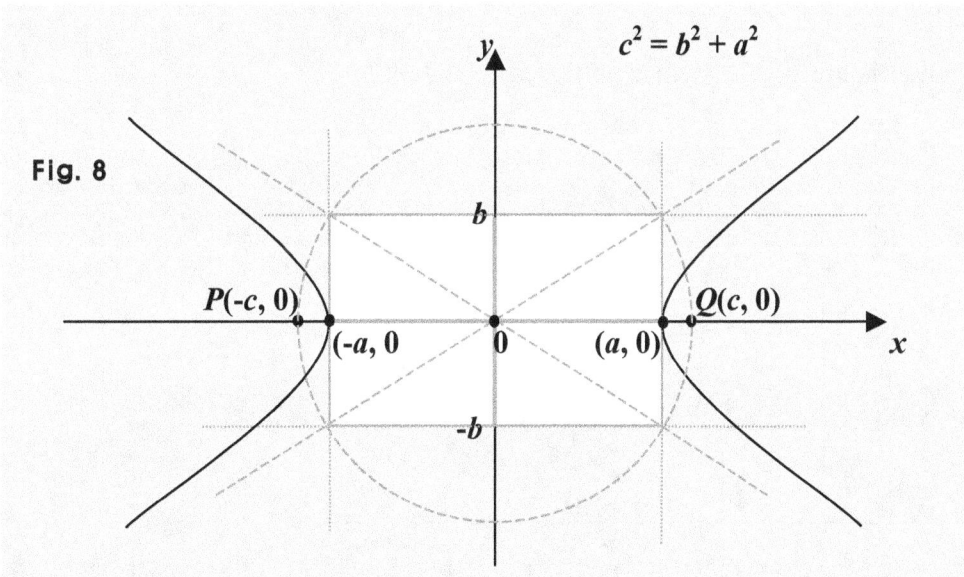

The other is: $\dfrac{y^2}{b^2} - \dfrac{x^2}{a^2} = 1$, where $c^2 = b^2 + a^2$, and c is the focal distance.

The hyperbola above is <u>vertical</u>, the <u>center is at (0, 0)</u>, the <u>transverse axis is **2b**</u>, and is a part of (thus, parallel to) the **y**-axis, and the conjugate axis is **2a**.
So the hyperbola is as below:

46

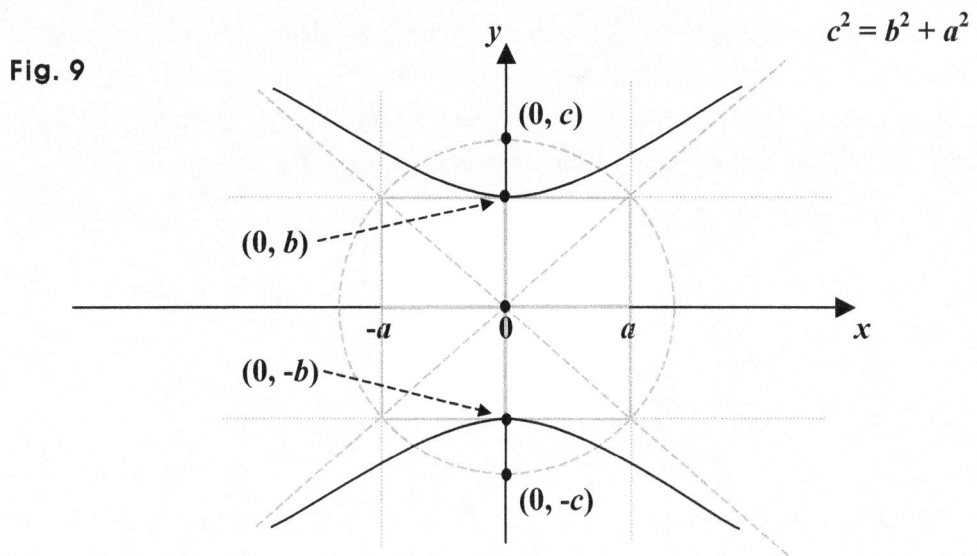

Fig. 9

$c^2 = b^2 + a^2$

What if the center is not at the origin as in the case below?

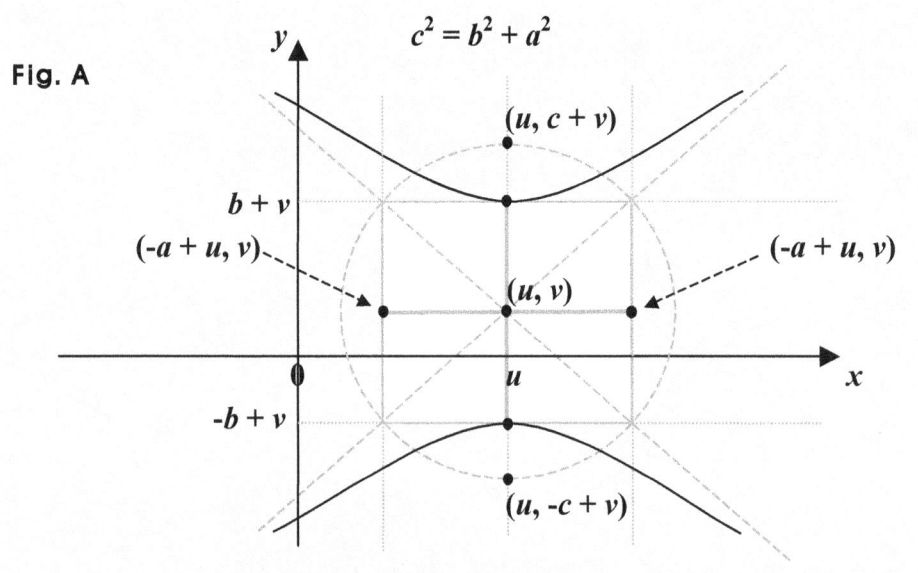

Fig. A

$c^2 = b^2 + a^2$

Then, the equation is: $\dfrac{(y-v)^2}{b^2} - \dfrac{(x-u)^2}{a^2} = 1$, where $c^2 = b^2 + a^2$.

Note that we can get the hyperbola above translating a hyperbola in amount of *u* along the *x*-axis, and in the amount of *v* along the *y*-axis.

And the hyperbola to be translated is: $\dfrac{y^2}{b^2} - \dfrac{x^2}{a^2} = 1$.

And in the case below:

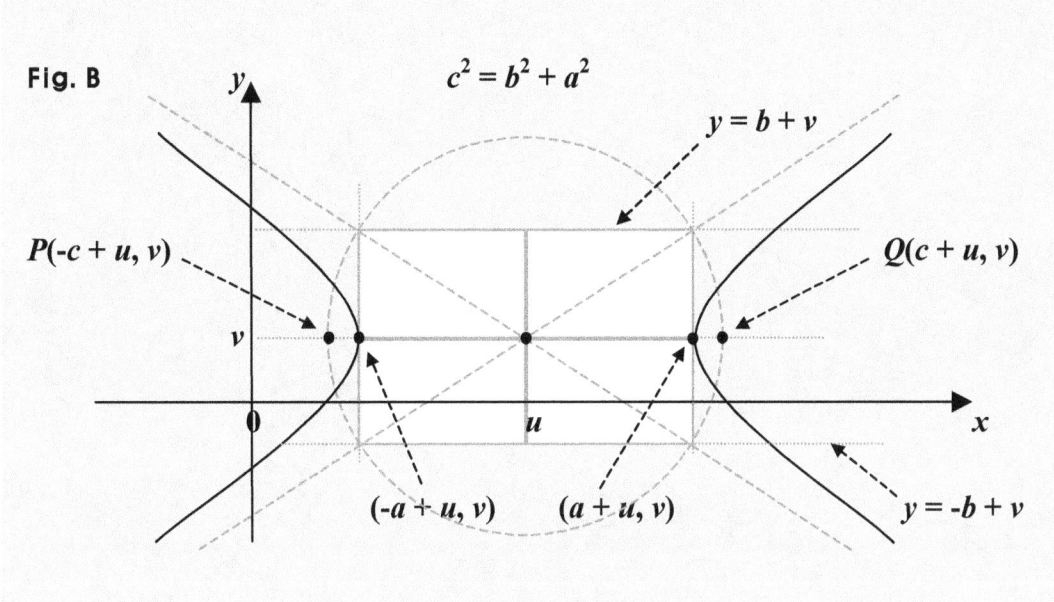

Fig. B

$c^2 = b^2 + a^2$

$y = b + v$

$P(-c + u, v)$

$Q(c + u, v)$

$(-a + u, v)$ $(a + u, v)$

$y = -b + v$

The equation is: $\dfrac{(x-u)^2}{a^2} - \dfrac{(y-v)^2}{b^2} = 1$, where $c^2 = b^2 + a^2$.

And note also that assuming **K** is the hyperbola above, we can get **K** translating a hyperbola in amount of *u* along the *x*-axis, and in the amount of *v* along the *y*-axis.

And the hyperbola to be translated is: $\dfrac{x^2}{a^2} - \dfrac{y^2}{b^2} = 1$.

And also, assuming **L** is the hyperbola above, we can get **L** translating the hyperbola **K** in amount of *–u* along the *x*-axis, and in the amount of *–v* along the *y*-axis.

Examples 3 in Standard Forms

Specify the center and the two main axes of each hyperbola below, and then, graph it.

0. $\dfrac{(x+3)^2}{9} - (y-5)^2 = 1$ 1. $\dfrac{(x+1)^2}{9} - \dfrac{(y-5)^2}{4} = 1$ 2. $(x-6)^2 - 4(y-\tfrac{7}{2})^2 = 9$

3. $4(x+5)^2 - 9y^2 = 36$ 4. $9x^2 - 16y^2 = 144$ 5. $4(x-3)^2 - 9(y+4)^2 = 36$

6. $9(x-7)^2 - 16(y+\tfrac{5}{2})^2 = 36$ 7. $(x+4)^2 - 4(y+\tfrac{11}{2})^2 = 9$

Fig. 0

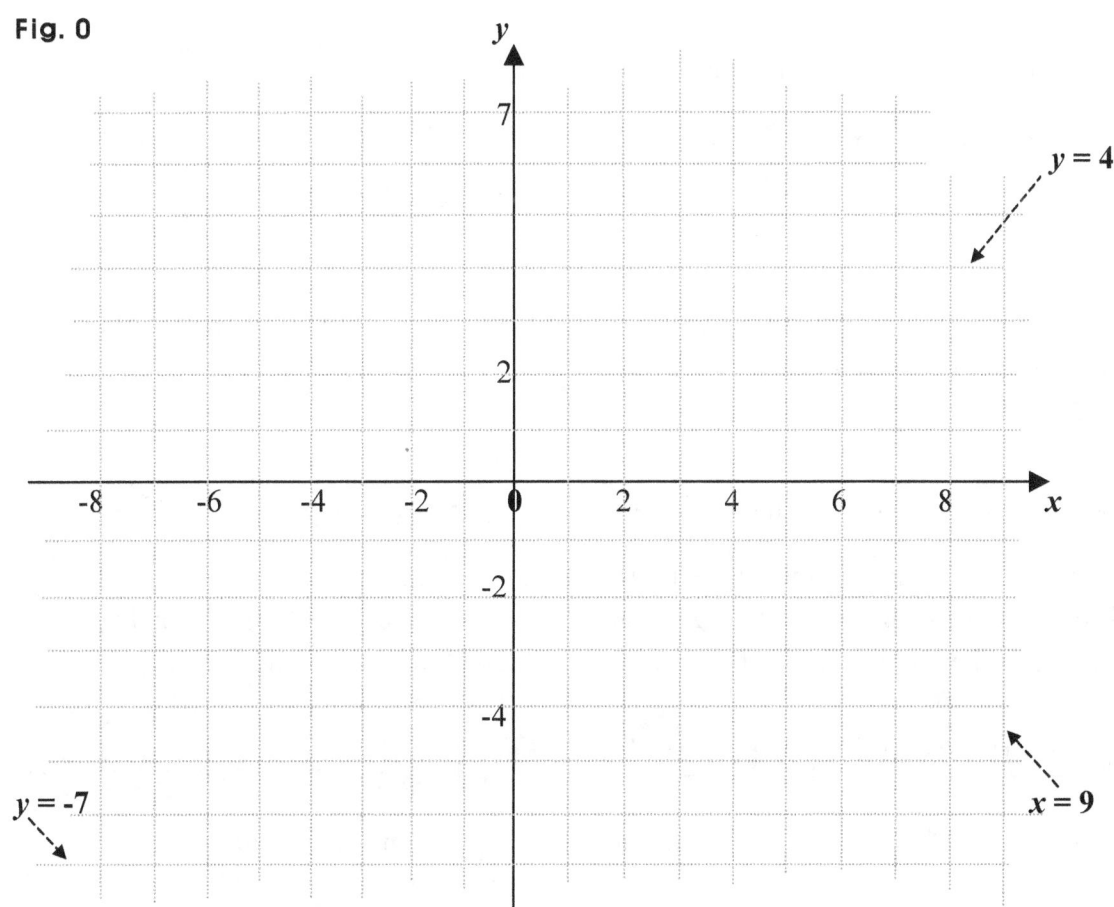

Suggestions or Solutions
To the Problems in the Examples

Specify the center, the transverse axis, and the conjugate axis of each hyperbola below, and then, graph it.

0. $\dfrac{(x+3)^2}{9}-(y-5)^2=1$ 1. $\dfrac{(x+1)^2}{9}-\dfrac{(y-5)^2}{4}=1$ 2. $(x-6)^2-4(y-\tfrac{7}{2})^2=9$

3. $4(x+5)^2-9y^2=36$ 4. $9x^2-16y^2=144$ 5. $4(x-3)^2-9(y+4)^2=36$

6. $9(x-7)^2-16(y+\tfrac{5}{2})^2=36$ 7. $(x+4)^2-4(y+\tfrac{11}{2})^2=9$

First, we have a fact as follows:

If a hyperbola centered at (**u, v**) is <u>horizontal</u>, the equation is: $\dfrac{(x-u)^2}{a^2}-\dfrac{(y-v)^2}{b^2}=1$, where **a** <u>is half the transverse axis</u>, and **b** is half the conjugate axis, and if <u>vertical</u>, the equation is: $\dfrac{(y-v)^2}{b^2}-\dfrac{(x-u)^2}{a^2}=1$, where **b** <u>is half the transverse axis</u>, and **a** is half the conjugate axis.

In either case though, the rectangle is **a** by **b**, **a** is the width (horizontal length) of the rectangle, and **b** is the height (vertical length).

So beginning with 0, we can get: $\dfrac{(x+3)^2}{9}-(y-5)^2=1 \Rightarrow \dfrac{\{x-(-3)\}^2}{3^3}-\dfrac{(y-5)^2}{1^2}=1$.

So we can see that the hyperbola is horizontal, the center is at (-3, 5), the transverse axis is 6, and the conjugate axis is 2. And its rectangle is centered at (-3, 5), too, and is 6 by 2, that is, the horizontal length of the rectangle is 6, and the vertical length is 2.

1. $\dfrac{(x+1)^2}{9}-\dfrac{(y-5)^2}{4}=1 \Rightarrow \dfrac{\{x-(-1)\}^2}{3^2}-\dfrac{(y-5)^2}{2^2}=1$. So we can see that the hyperbola is horizontal, the center is (-1, 5), the transverse axis is 6, and the conjugate axis is 4. And the rectangle is 6 by 4, that is, the width is 6, and the height is 4.

2. $(x-6)^2 - 4(y-\frac{7}{2})^2 = 9 \Rightarrow \dfrac{(x-6)^2}{9} - \dfrac{4(y-\frac{7}{2})^2}{9} = 1 \Rightarrow \dfrac{(x-6)^2}{3^2} - \dfrac{(y-\frac{7}{2})^2}{(\frac{3}{2})^2} = 1.$

So the hyperbola is horizontal, the center is (-1, $\frac{7}{2}$), the transverse axis is 6, and the conjugate axis is 3. And the rectangle is 6 by 3, that is, the width is 6, and the height is 3.

3. $4(x+5)^2 - 9y^2 = 36 \Rightarrow \dfrac{4(x+5)^2}{36} - \dfrac{9y^2}{36} = 1 \Rightarrow \dfrac{(x+5)^2}{9} - \dfrac{y^2}{4} = 1 \Rightarrow \dfrac{(x+5)^2}{3^2} - \dfrac{y^2}{2^2} = 1.$

So the hyperbola is horizontal, the center is (-5, 0), the transverse axis is 6, and the conjugate axis is 4. And the rectangle is 6 by 4, that is, the width is 6, and the height is 4.

4. $9x^2 - 16y^2 = 144 \Rightarrow \dfrac{9x^2}{144} - \dfrac{16y^2}{144} = \dfrac{x^2}{16} - \dfrac{y^2}{9} = \dfrac{x^2}{4^2} - \dfrac{y^2}{3^2} = 1.$

So the hyperbola is horizontal, the center is (0, 0), the transverse axis is 8, and the conjugate axis is 6. And the rectangle is 8 by 6, that is, the width is 8, and the height is 6.

5. $4(x-3)^2 - 9(y+4)^2 = 36 \Rightarrow \dfrac{(x-3)^2}{9} - \dfrac{(y+4)^2}{4} = \dfrac{(x-3)^2}{3^2} - \dfrac{(y+4)^2}{2^2} = 1.$

So the hyperbola is horizontal, the center is (3, -4), the transverse axis is 6, and the conjugate axis is 4. And the rectangle is 6 by 4, that is, the width is 6, and the height is 4.

6. $9(x-7)^2 - 16(y+\frac{5}{2})^2 = 36 \Rightarrow \dfrac{(x-7)^2}{4} - \dfrac{4(y+\frac{5}{2})^2}{9} = \dfrac{(x-7)^2}{2^2} - \dfrac{(y+\frac{5}{2})^2}{(\frac{3}{2})^2} = 1.$

So the hyperbola is horizontal, the center is (7, $-\frac{5}{2}$), the transverse axis is 4, and the conjugate axis is 3. And the rectangle is 4 by 3, that is, the width is 4, and the height is 3.

7. $(x+4)^2 - 4(y+\frac{11}{2})^2 = 9 \Rightarrow \dfrac{(x+4)^2}{9} - \dfrac{4(y+\frac{11}{2})^2}{9} = \dfrac{(x+4)^2}{3^2} - \dfrac{(y+\frac{11}{2})^2}{(\frac{3}{2})^2} = 1.$

So the hyperbola is horizontal, the center is ($-4, -\frac{11}{2}$), the transverse axis is 6, and the conjugate axis is 3. And the rectangle is 6 by 3, that is, the width is 6, and the height is 3.

And graphing the equations, we get:

Fig. 1

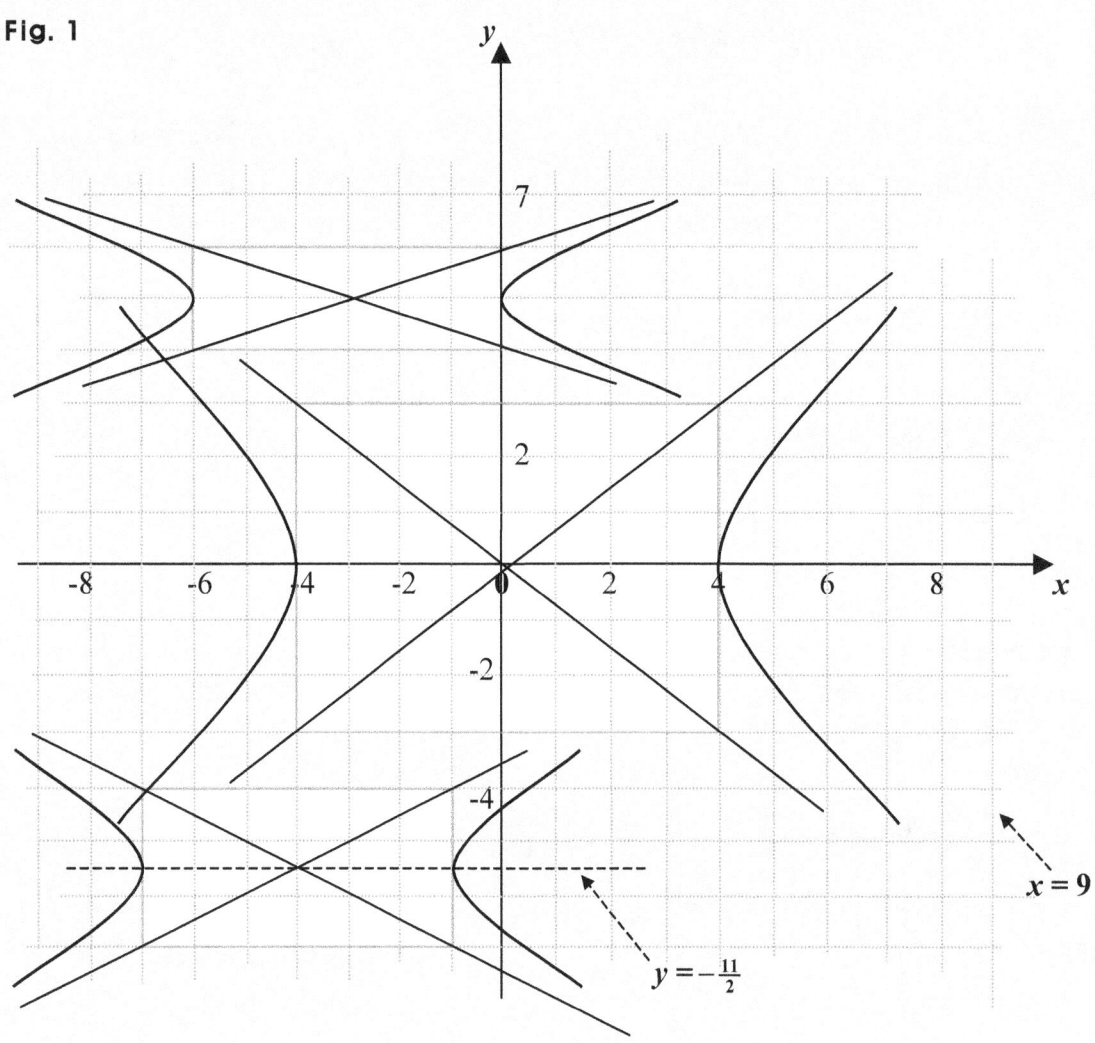

There are more hyperbolas in the next page.

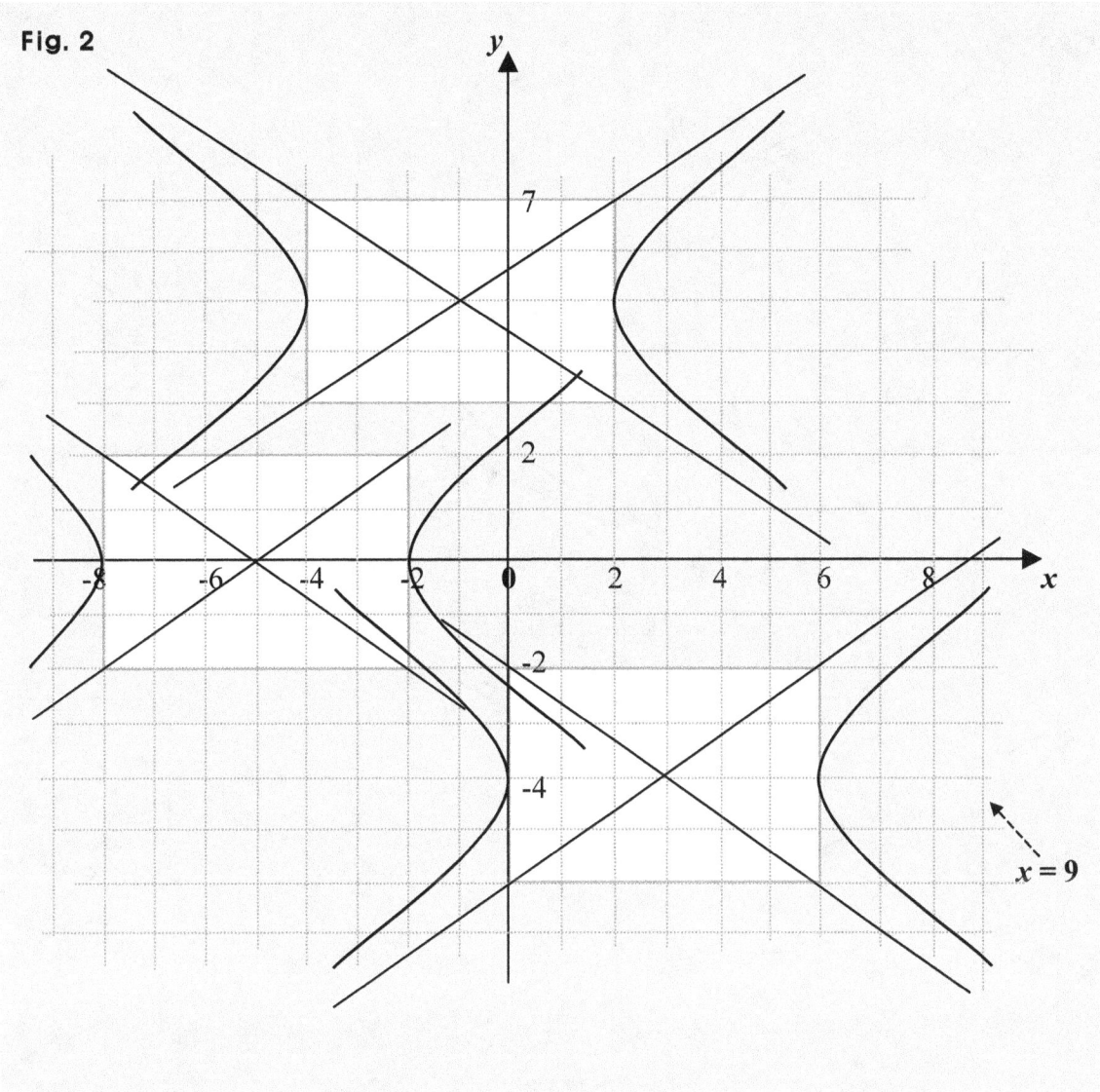

All the hyperbolas themselves above are actually the same, but have different positions in the graph, so they are considered to be different hyperbolas, and thus in math, have different equations.

And the next page has the other hyperbolas.

54

Fig. 3

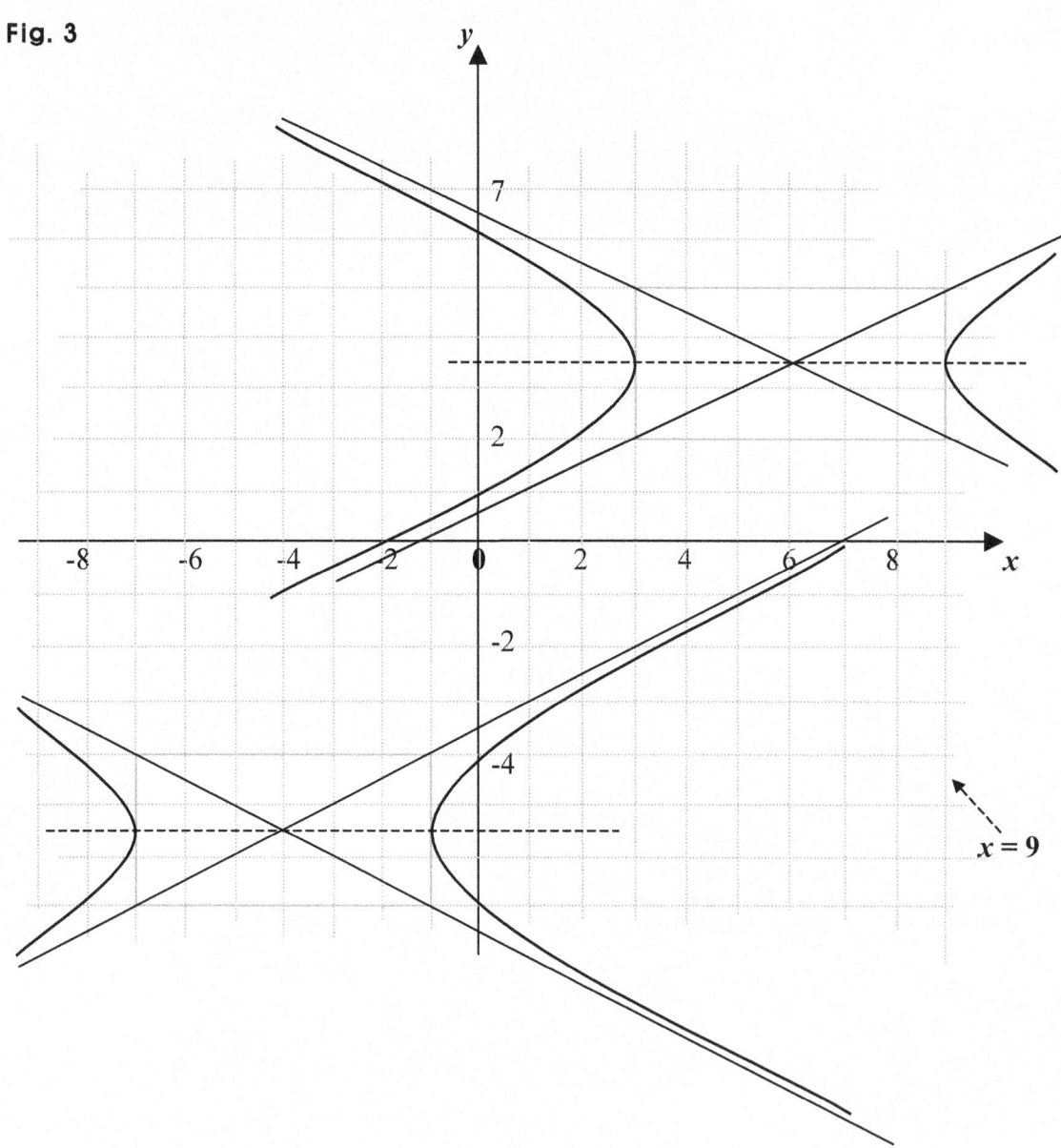

Hyperbolas are symmetric about its center, the conjugate axis, and the transverse axis, so each hyperbola above is shown just in part and not as a whole.

In fact, we cannot show a hyperbola entirely, because its length is infinite.

And although the hyperbola looks meeting the asymptotes as it gets longer, it does not meet them no matter how long it may get longer, and that's what an asymptote means.

Examples 4 in Standard Forms

Specify the center and the two main axes of each hyperbola below, and then, graph it.

0. $4(y-5)^2 - (x+4)^2 = 4$ 1. $4y^2 - x^2 = 4$ 2. $4(y+\frac{11}{2})^2 - 9(x+2)^2 = 9$

3. $9(y-5)^2 - 4(x+1)^2 = 36$ 4. $(y+3)^2 - (x+6)^2 = 1$ 5. $(y+4)^2 - (x-4)^2 = 4$

6. $400(y+\frac{11}{2})^2 - 36(x+\frac{9}{2})^2 = 225$ 7. $144(y-\frac{5}{2})^2 - 100(x-\frac{9}{2})^2 = 225$

Fig. 0

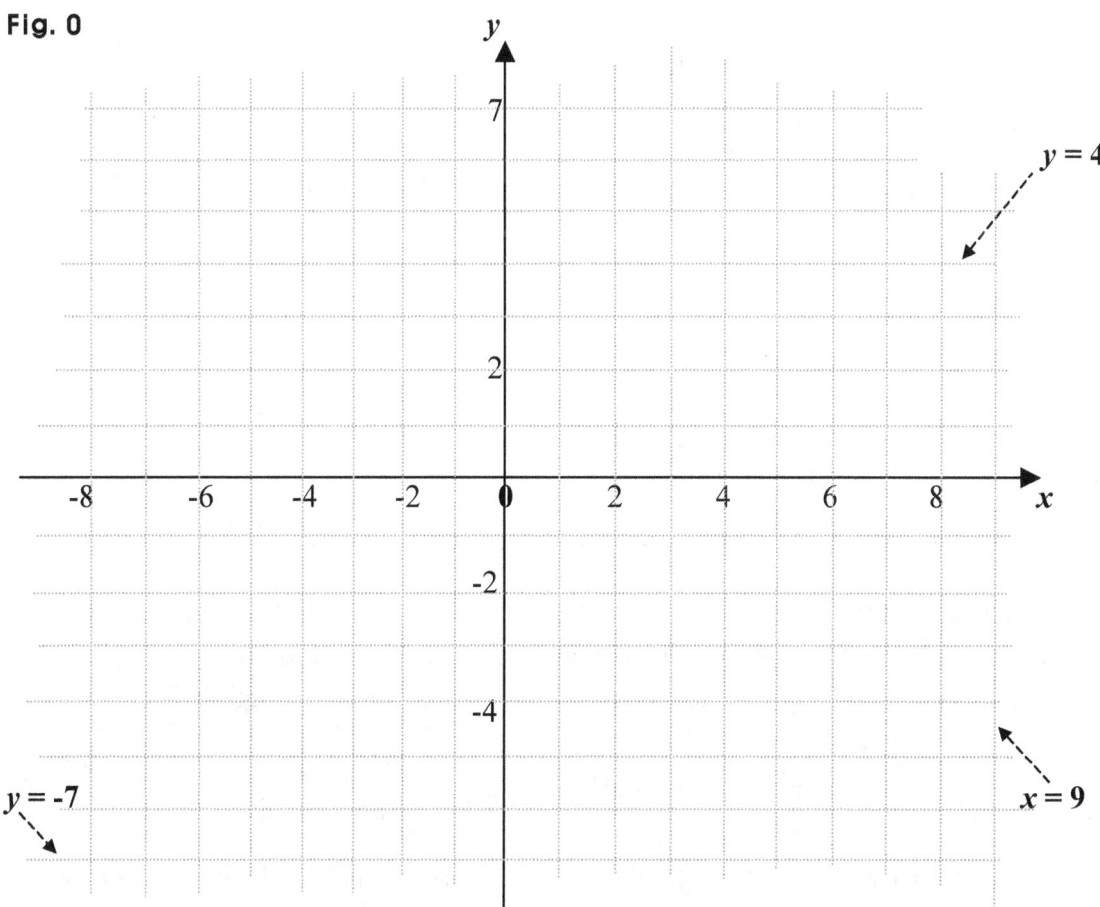

Suggestions or Solutions
To the Problems in the Examples

Specify the center, the transverse axis, and the conjugate axis of each hyperbola below, and then, graph it.

0. $4(y-5)^2 - (x+4)^2 = 4$ 1. $4y^2 - x^2 = 4$ 2. $4(y+\frac{11}{2})^2 - 9(x+2)^2 = 9$

3. $9(y-5)^2 - 4(x+1)^2 = 36$ 4. $(y+3)^2 - (x+6)^2 = 1$ 5. $(y+4)^2 - (x-4)^2 = 4$

6. $400(y+\frac{11}{2})^2 - 36(x+\frac{9}{2})^2 = 225$ 7. $144(y-\frac{5}{2})^2 - 100(x-\frac{9}{2})^2 = 225$

First, we have a fact as follows:

If a hyperbola centered at (u, v) is <u>horizontal</u>, the equation is: $\dfrac{(x-u)^2}{a^2} - \dfrac{(y-v)^2}{b^2} = 1$, where a is half the transverse axis, and b is half the conjugate axis, and if <u>vertical</u>, the equation is: $\dfrac{(y-v)^2}{b^2} - \dfrac{(x-u)^2}{a^2} = 1$, where $\underline{b}$ is half the transverse axis, and a is half the conjugate axis.

In either case though, the rectangle is a by b, a is the width (horizontal length) of the rectangle, and b is the height (vertical length).

So beginning with 0, we can get:

$$4(y-5)^2 - (x+4)^2 = 4 \Rightarrow (y-5)^2 - \frac{(x+4)^2}{4} = 1 \Rightarrow \frac{(y-5)^2}{1^2} - \frac{\{x-(-4)\}^2}{2^2} = 1.$$

So we can see that the hyperbola is vertical, the center is at (-4, 5), the transverse axis is 2, and the conjugate axis is 4. And its rectangle is centered at (-4, 6), too, and is 4 by 2, that is, the horizontal length of the rectangle is 4, and the vertical length is 2.

1. $4y^2 - x^2 = 4 \Rightarrow y^2 - \frac{x^2}{4^2} = 1 \Rightarrow \frac{(y-0)^2}{1^2} - \frac{(x-0)^2}{4^2} = 1.$ So we can see that the

hyperbola is vertical, the center is (0, 0), the transverse axis is 2, and the conjugate axis is 8. And the rectangle is 8 by 2, that is, the width is 8, and the height is 2.

2. $4(y+\frac{11}{2})^2 - 9(x+2)^2 = 9 \Rightarrow \dfrac{4(y+\frac{11}{2})^2}{9} - (x+2)^2 = 1$

$\Rightarrow \dfrac{4(y+\frac{11}{2})^2}{9} - (x+2)^2 = \dfrac{(y+\frac{11}{2})^2}{\frac{9}{4}} - (x+2)^2 = \dfrac{(y+\frac{11}{2})^2}{(\frac{3}{2})^2} - \dfrac{(x+2)^2}{1^2} = 1.$

So the hyperbola is vertical, the center is $(-2, -\frac{11}{2})$, the transverse axis is 3, and the conjugate axis is 2. And the rectangle is 2 by 3, that is, the width is 2, and the height is 3.

3. $9(y-5)^2 - 4(x+1)^2 = 36 \Rightarrow \dfrac{(y-5)^2}{4} - \dfrac{(x+1)^2}{9} = 1$

$\Rightarrow \dfrac{(y-5)^2}{4} - \dfrac{(x+1)^2}{9} = \dfrac{(y-5)^2}{2^2} - \dfrac{(x+1)^2}{3^2} = 1.$

So the hyperbola is vertical, the center is (-1, 5), the transverse axis is 4, and the conjugate axis is 6. And the rectangle is 6 by 4, that is, the width is 6, and the height is 4.

4. $(y+3)^2 - (x+6)^2 = 1 \Rightarrow \dfrac{(y+3)^2}{1^2} - \dfrac{(x+6)^2}{1^2} = 1.$

So the hyperbola is vertical, the center is (-6, -3), the transverse axis is 2, and the conjugate axis is 2. And the rectangle is a 2 by 2 square.

5. $(y+4)^2 - (x-4)^2 = 4 \Rightarrow \dfrac{(y+4)^2}{2^2} - \dfrac{(x-4)^2}{2^2} = 1.$

So the hyperbola is vertical, the center is (4, -4), the transverse axis is 4, and the conjugate axis is 4. And the rectangle is a 4 by 4 square.

6. $400(y+\frac{11}{2})^2 - 36(x+\frac{9}{2})^2 = 225 \Rightarrow \dfrac{16(y+\frac{11}{2})^2}{9} - \dfrac{4(x+\frac{9}{2})^2}{25} = 1$

$\Rightarrow \dfrac{16(y+\frac{11}{2})^2}{9} - \dfrac{4(x+\frac{9}{2})^2}{25} = \dfrac{4(y+\frac{11}{2})^2}{\frac{9}{4}} - \dfrac{(x+\frac{9}{2})^2}{\frac{25}{4}} = \dfrac{4(y+\frac{11}{2})^2}{(\frac{3}{2})^2} - \dfrac{(x+\frac{9}{2})^2}{(\frac{5}{2})^2} = 1$

So the hyperbola is vertical, the center is $(-\frac{9}{2}, -\frac{11}{2})$, the transverse axis is 3, and the conjugate axis is 5. And the rectangle is 5 by 3, that is, the width is 5, and the height is 3.

7. $144(y-\frac{5}{2})^2 - 100(x-\frac{9}{2})^2 = 225 \Rightarrow \dfrac{16(y-\frac{5}{2})^2}{25} - \dfrac{4(x-\frac{9}{2})^2}{9} = 1$

$\Rightarrow \dfrac{16(y-\frac{5}{2})^2}{25} - \dfrac{4(x-\frac{9}{2})^2}{9} = \dfrac{4(y-\frac{5}{2})^2}{\frac{25}{4}} - \dfrac{(x-\frac{9}{2})^2}{\frac{9}{4}} = \dfrac{4(y-\frac{5}{2})^2}{(\frac{5}{2})^2} - \dfrac{(x-\frac{9}{2})^2}{(\frac{3}{2})^2} = 1$

So the hyperbola is vertical, the center is $(\frac{9}{2}, \frac{5}{2})$, the transverse axis is 5, and the conjugate axis is 3. And the rectangle is 3 by 5, that is, the width is 3, and the height is 5.

And graphing the equations, we get:

Fig. 1

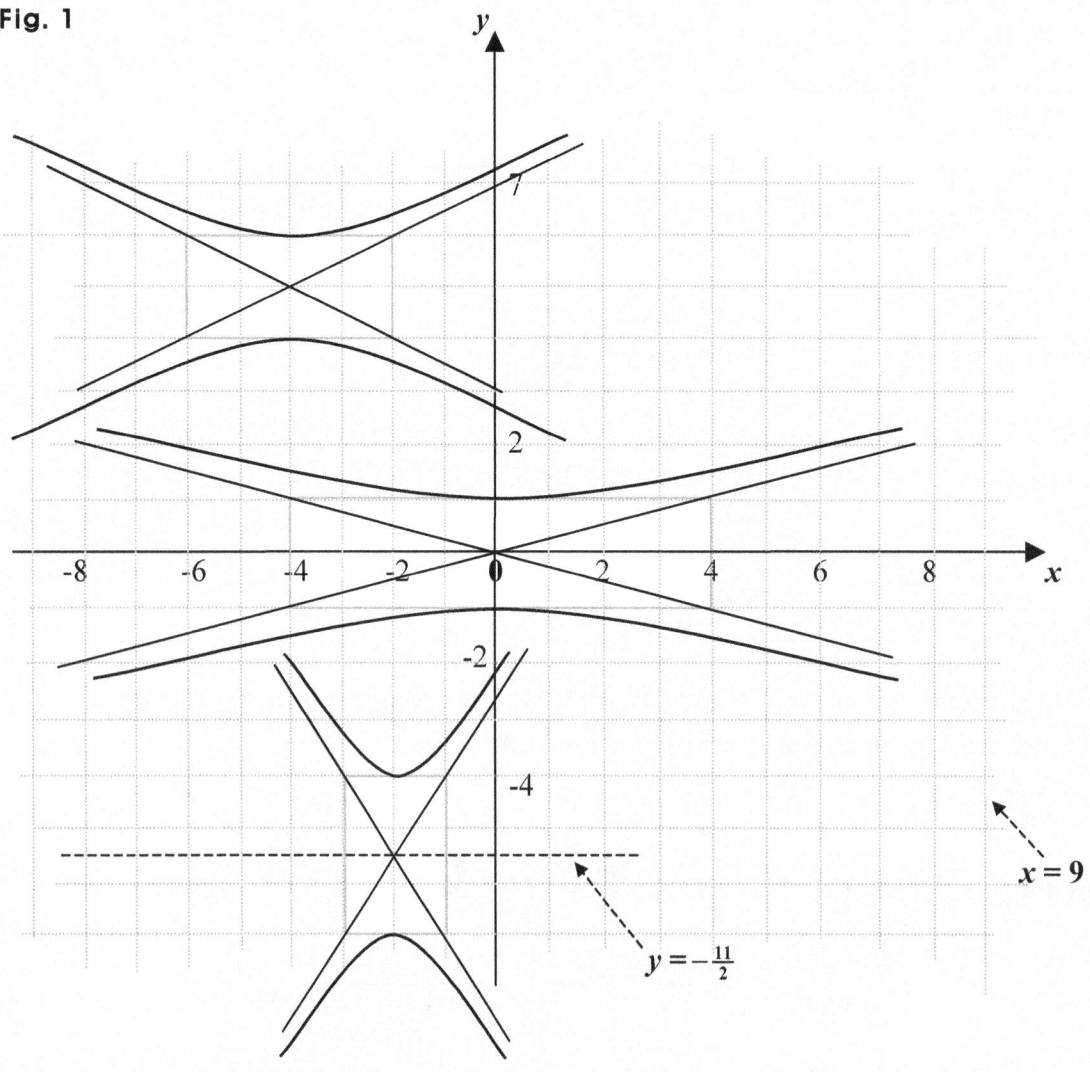

There are more hyperbolas in the next page.

Fig. 2

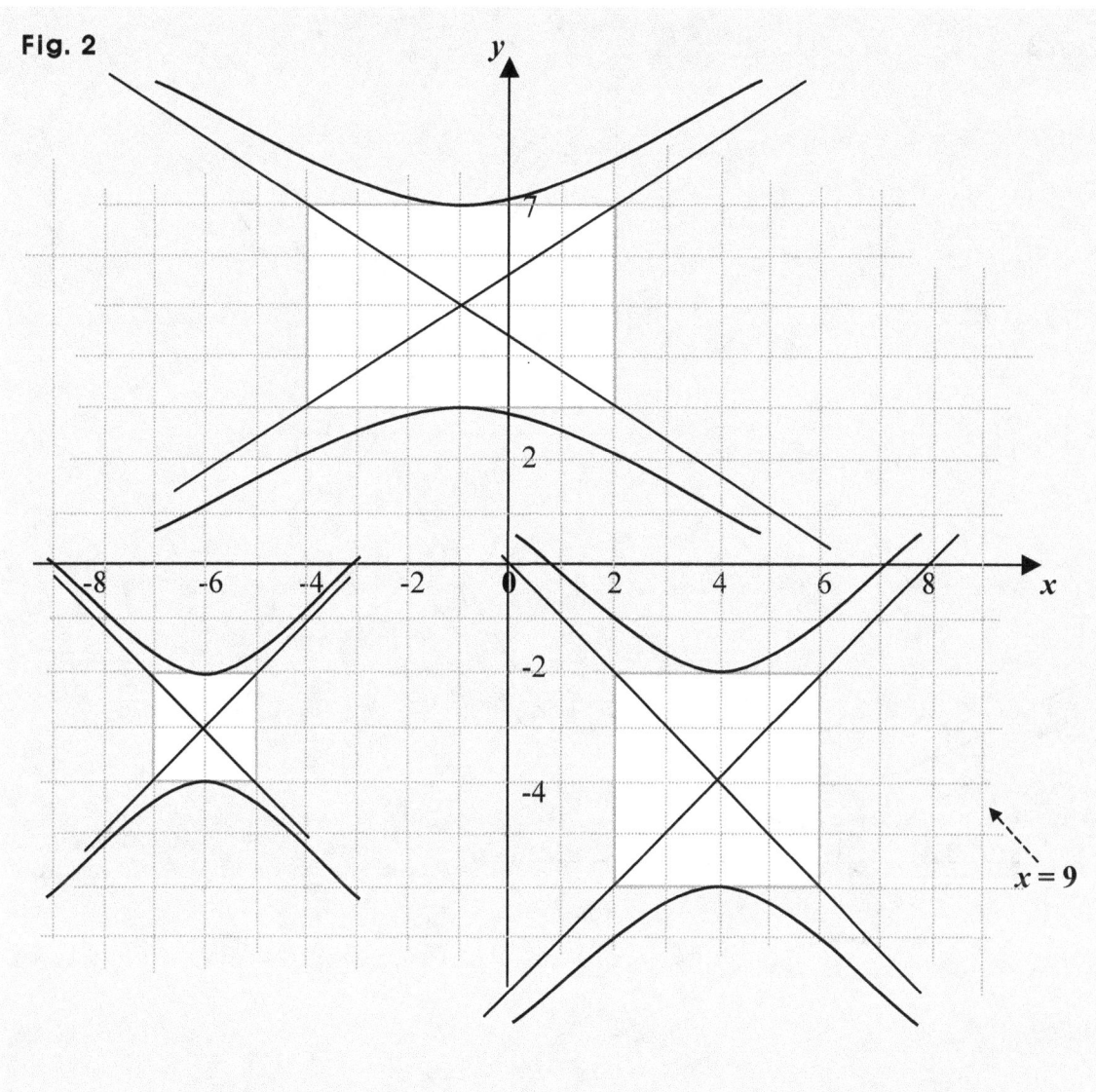

And the next page shows the other hyperbolas.

Fig. 3

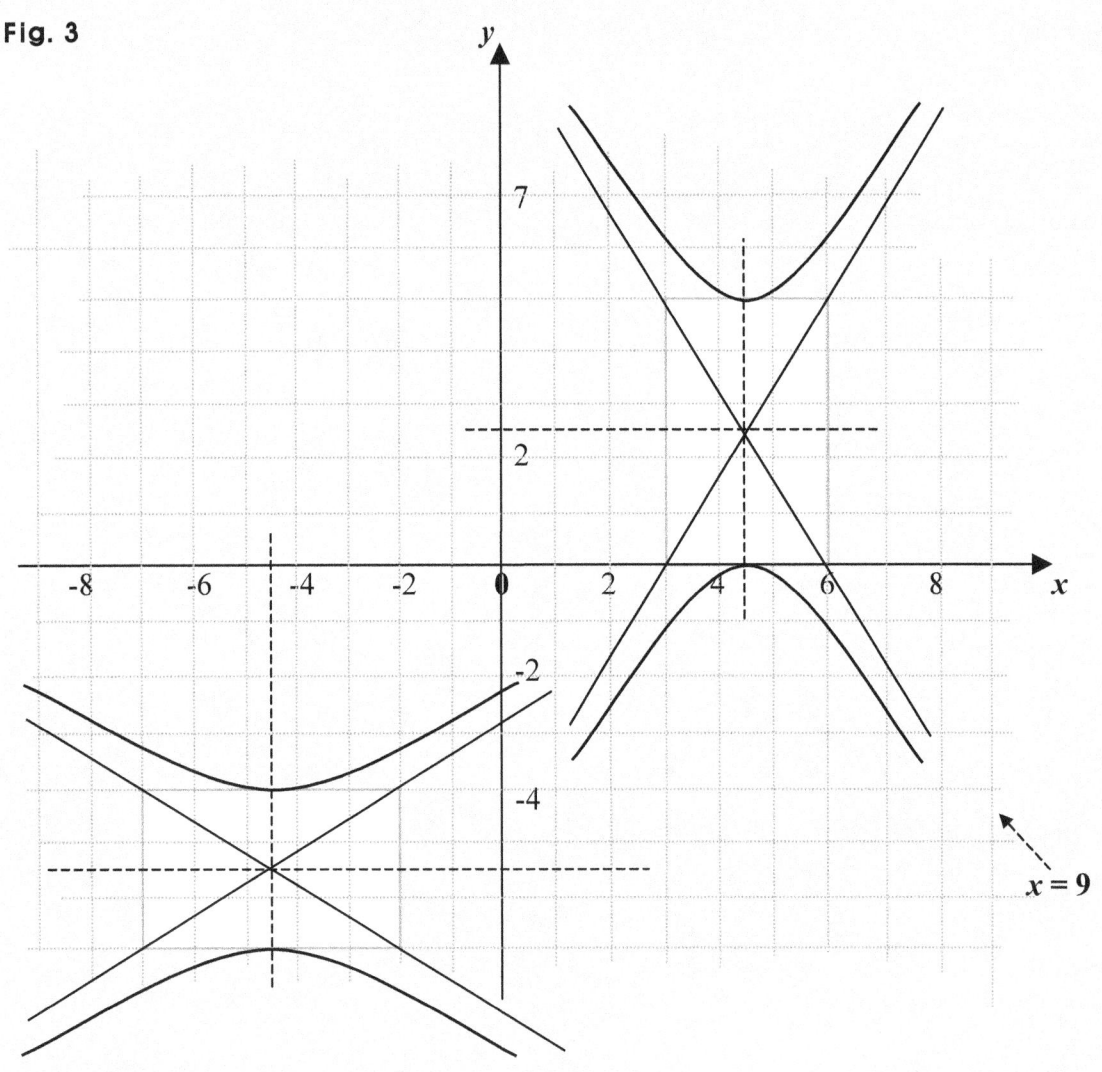

3. Elements of a Hyperbola

As in the case of an ellipse, working with a hyperbola, too, we often need to refer to some basic elements in it. And the elements are as follows:

center, foci, directrices, vertices, main axes, focal distance, eccentricity, and asymptotes.

So let's now take a close look at each of those elements.

To begin with, a hyperbola has <u>two vertices</u>, which are <u>both ends of the transverse axis</u>.

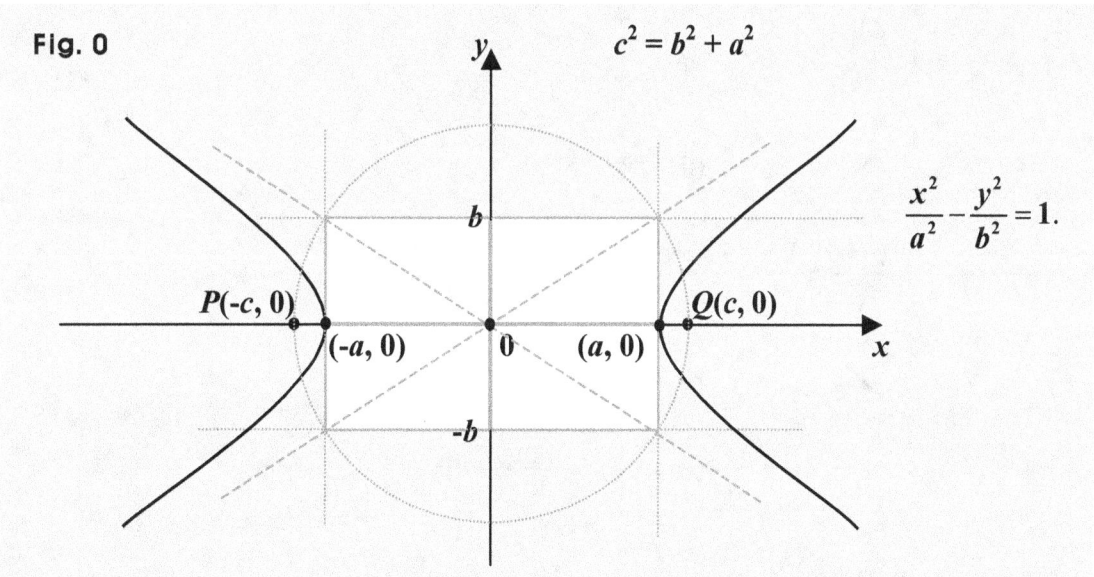

Fig. 0

$$c^2 = b^2 + a^2$$

$$\frac{x^2}{a^2} - \frac{y^2}{b^2} = 1.$$

$P(-c, 0)$ $Q(c, 0)$

$(-a, 0)$ 0 $(a, 0)$

b $-b$

So in the hyperbola above, the two vertices are **(-a, 0)** and **(a, 0)**.   What about **P** and **Q**?

The two points **P(-c, 0)** and **Q(c, 0)** are the foci, and the center is **(0, 0)**.

And the transverse axis is a part of a line called the transverse line, and we can notice that <u>the transverse line has all the five points</u>: the vertices, the foci, and the center.

So we can also notice that if a hyperbola is <u>horizontal, the y-coordinates are the same</u> at all those five points, which share thus, the same y-coordinate.

Then, we can notice also that if a hyperbola is <u>vertical</u>, the *x*-coordinates are the same at all the five points: the vertices, foci, and center, which share thus, the same *x*-coordinate.

• Next, a hyperbola is symmetric, and has two axes of symmetry, called main axes, too.

One of the two is called the major axis, and the other is called the minor axis. Usually though, <u>the major is called the transverse axis</u>, and the minor is called the conjugate axis.

And in the case of the hyperbola below, the line segment connecting the vertices is the transverse axis, the length of which is **2a**, and the line segment connecting the two points **(0, b)** and **(0, -b)** is the conjugate axis, the length of which is **2b**. Usually though, we just call **2a** the transverse axis, and call **2b** the conjugate axis.

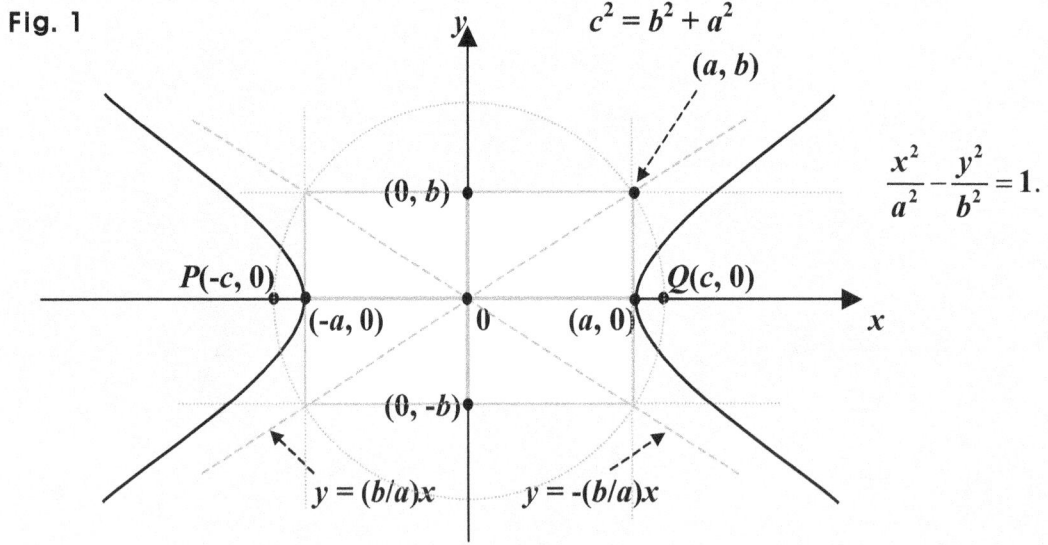

Fig. 1

$$c^2 = b^2 + a^2$$

$$\frac{x^2}{a^2} - \frac{y^2}{b^2} = 1.$$

What then, about *a* and *b*?

We call *a* the <u>semi major</u> axis, and *b* is called the <u>semi minor</u> axis.

And a hyperbola has two lines called <u>asymptotes</u>, each of which has a slope. We can get <u>the two slopes using the semis: *a* and *b*</u>. Note however, in cases of asymptotes, it doesn't matter which of *a* and *b* is the semi major axis. We just use the definition for slopes: rise over run. So in the case of the asymptotes above, one slope is: ***b*/*a***, and the other is: ***-b*/*a***. And since both asymptotes pass through the origin, they are the two lines ***y* = ±(*b*/*a*)*x***.

And as we can see in the equation of the hyperbola above, if a hyperbola is <u>horizontal</u>, <u>the sign</u> of x^2-term is <u>positive</u>, and the sign of y^2-term is negative.

And if it is vertical, <u>the sign</u> of x^2-term is <u>negative</u>, and the sign of y^2-term is positive.

• Next, the focal distance is the distance from a focus to the center.

The two foci are symmetric about the center. So the focal distance is half the distance between the foci. So in the hyperbola above, <u>c is the focal distance</u>.

• Next, in the case of the hyperbola above, the two foci are **(-c, 0)** and **(c, 0)**.

The difference between two distances from every point in a hyperbola to its two foci is constant, that is, the same. And in the case of the hyperbola above, <u>the difference is **2a**</u>, which is (the length of) <u>the transverse axis</u>.

What then, about this: $c^2 = b^2 + a^2$?

It can be called <u>the connective equation between the focal distance and the semi axes</u>. And we can get the equation from the three facts as follows:

• One is that c is the focal distance, that is, the distance from one focus to the center.

• Another is that the focal distance is the radius of the circle that is centered at the center of the hyperbola, and is passing through the foci of the hyperbola.

• And the other fact is that a right triangle is made by three points, one is <u>a point where the circle meets an asymptote</u>, another is <u>a vertex</u>, and the other is <u>the center</u>.

And in the case of the hyperbola above, the three points can be: **(a, b)**, **(a, 0)**, and **(0, 0)**.

So the three points above determine a right triangle, where the hypotenuse is c, which is the focal distance. Using thus, the distance formula, we can get: $c^2 = b^2 + a^2$.

So the focal distance is: $c = \sqrt{a^2 + b^2}$.

And in the case of the horizontal hyperbola above, the two foci are: **(-c, 0)** and **(c, 0)**.

So we can put them this way, too: $(-\sqrt{a^2 + b^2},\ \mathbf{0})$ and $(\sqrt{a^2 + b^2},\ \mathbf{0})$.

What if a hyperbola is <u>vertical</u>, and the semis are **a** and **b**?

We still get: $c^2 = b^2 + a^2$.   So we get: $c = \sqrt{a^2 + b^2}$, too.

Therefore, in the case of a <u>vertical</u> hyperbola, too, the focal distance is: $c = \sqrt{a^2 + b^2}$.

And in the case of a hyperbola <u>vertical and centered at the origin</u>, the two foci are: **(0, c)** and **(0, -c)**, because the transverse line is the *y*-axis.

So in that case, we can put the foci this way, too: $(0,\ \sqrt{a^2 + b^2})$ and $(0,\ -\sqrt{a^2 + b^2})$.

• Next, a hyperbola has a special number called an <u>eccentricity</u>, which is a ratio of a focal distance to the semi major, i.e., <u>the focal distance over half the transverse axis.</u>

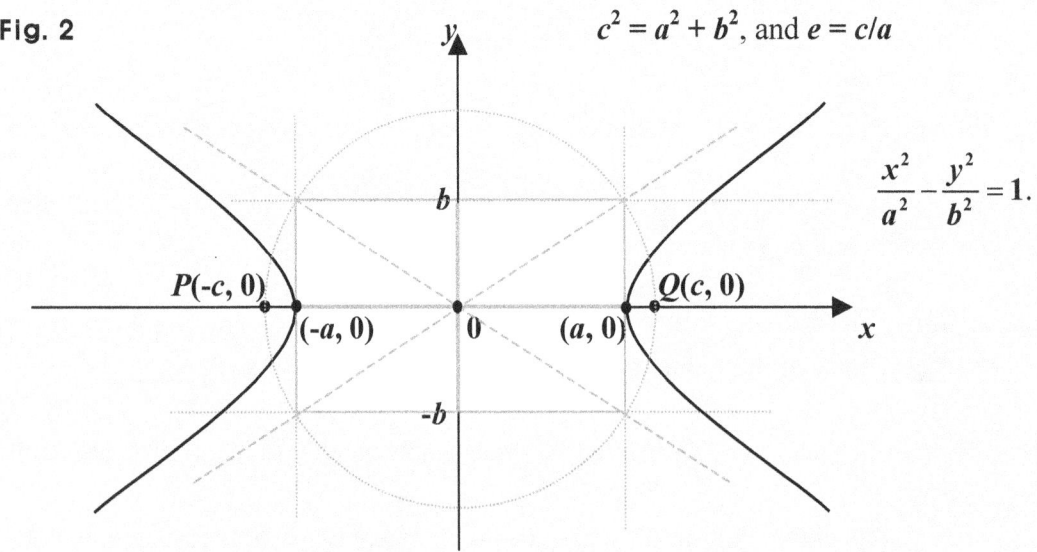

Fig. 2 $c^2 = a^2 + b^2$, and $e = c/a$

$$\frac{x^2}{a^2} - \frac{y^2}{b^2} = 1.$$

So of a hyperbola <u>horizontal</u>, the eccentricity is: $e = c/a$, and specifies the degree, to which the hyperbola bends. It can tell us thus, how flat or sharp the branches are.

Usually, an eccentricity is denoted by *e*, and we have: *e* > **1** for a hyperbola.

And if the eccentricity *e* were 1, each branch would be a ray.

So if the eccentricity is close to 1, each branch is very sharp at the vertex.

And if *e* is very large, the branches are very flat, and <u>look like</u> two parallel lines.

So the bigger the eccentricity, the flatter the branches.

Now, the eccentricity *e* is a ratio, which is <u>the focal distance over half the transverse</u>.

So in the case of <u>a hyperbola horizontal, we get</u>: *e* = *c*/*a*. because the transverse is **2a**.

And in the case of <u>a hyperbola vertical, we get</u>: *e* = *c*/*b*, because the transverse is **2b**.

So we can put the focal distance *c* the ways below, too:

If horizontal, since: *e* = *c*/*a*, we get: *c* = *ae*, and if vertical, since: *e* = *c*/*b*, we get: *c* = *be*.

How come though, each branch would be a ray if it were the case where *e* = **1**?

If a hyperbola is horizontal, we have: *e* = *c*/*a*.   So we get: *c* = *a* if *e* = **1**.

And next, we have: $c = \sqrt{a^2 + b^2}$, so if *c* = *a*, we would get: *b* = **0**.

Thus, if *e* = **1**, the invisible rectangle would be a line segment of length **2a**.

So if it were the case where *e* = **1**, the hyperbola would be as below:

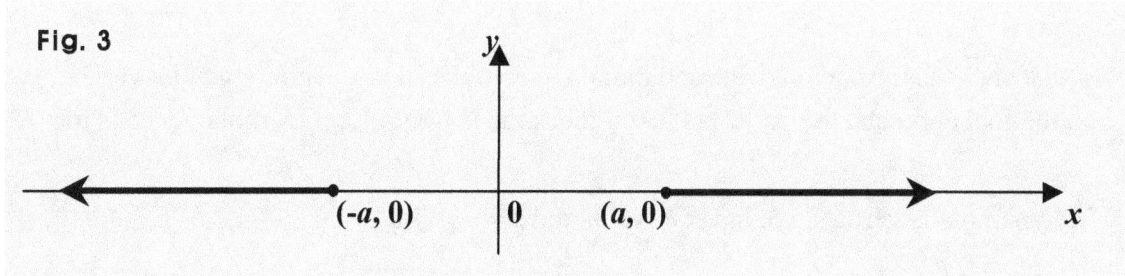

Fig. 3

Therefore, if *e* = **1**, each branch would be a ray beginning at the vertex.

Let's take another look at the eccentricity, but this time, a bit differently.

If a hyperbola is horizontal, we have: $e = c/a$, and $c = \sqrt{a^2 + b^2}$.

Then, we can put the eccentricity e this way, too: $e = \dfrac{\sqrt{a^2 + b^2}}{a} = \sqrt{\dfrac{a^2 + b^2}{a^2}} = \sqrt{1 + \dfrac{b^2}{a^2}}$.

Then, as b changes and approaches 0, the eccentricity e changes and approaches 1. And since: $c = \sqrt{a^2 + b^2}$, the focal distance c approaches a. So the foci approach the vertices. Thus, as b changes the way above, each branch changes and approaches a ray beginning at its vertex.

In the figure below, we can see that as b gets smaller, the branches get sharper at the vertices. So both branches would be two rays beginning at the vertices if b were 0.

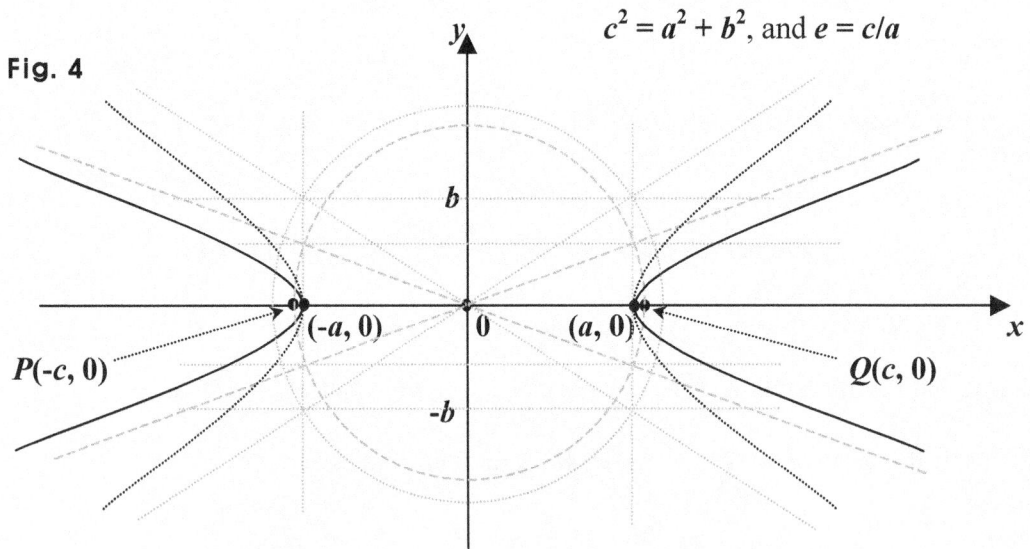

Fig. 4

So as b changes the way above, each branch approaches a ray beginning at its vertex, and the foci approach the vertices.   And the same is true for a hyperbola vertical, too

If a hyperbola is vertical, we have: $e = c/b$, and $c = \sqrt{a^2 + b^2}$.

Then, we can put the eccentricity e this way, too: $e = \dfrac{\sqrt{a^2 + b^2}}{b} = \sqrt{\dfrac{b^2 + a^2}{b^2}} = \sqrt{1 + \dfrac{a^2}{b^2}}$.

Then, as a changes and approaches 0, the eccentricity e changes and approaches 1. And since: $c = \sqrt{a^2 + b^2}$, the focal distance c approaches b. So the foci approach the vertices.

Thus, as a changes the way above, each branch changes and approaches a ray beginning at its vertex.

• And next, as in the case of an ellipse, a hyperbola has two **directrices**, too.

In the figure below, the two vertical lines in gray are the two directrices, one is with the left branch, and the other is with the right branch.

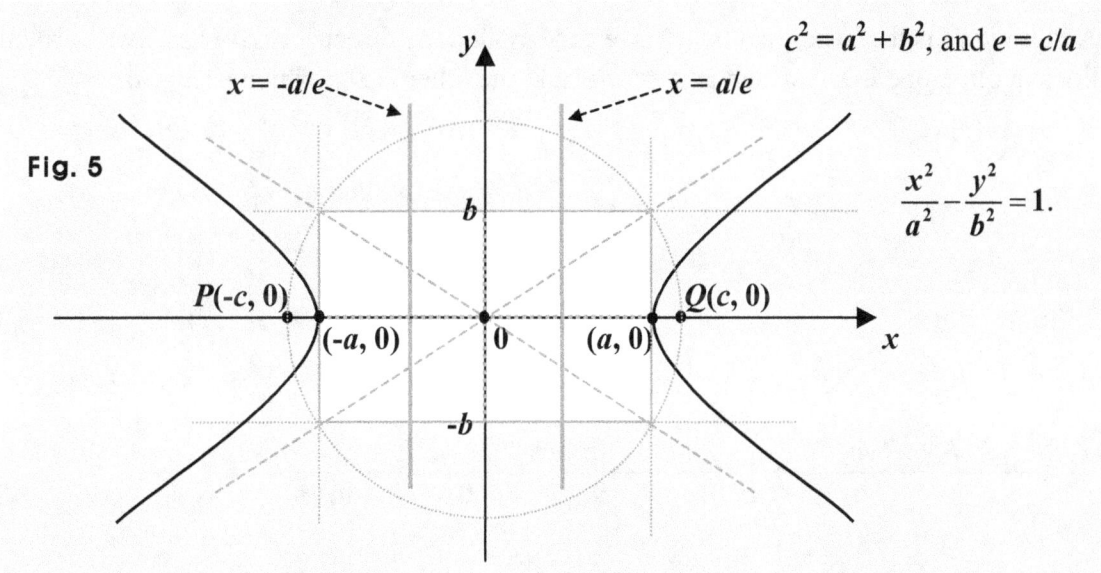

Fig. 5

$c^2 = a^2 + b^2$, and $e = c/a$

$x = -a/e$ $x = a/e$

$\dfrac{x^2}{a^2} - \dfrac{y^2}{b^2} = 1.$

$P(-c, 0)$ $Q(c, 0)$

$(-a, 0)$ 0 $(a, 0)$ x

So if a hyperbola is horizontal, directrices are vertical lines, that is, lines perpendicular to the x-axis. And the focus on the left is said to correspond to the directrix on the left, and the focus on the right is said to correspond to the directrix on the right.

So in the figure above, the vertical line $x = -a/e$ is the directrix corresponding to the focus P, and the other line $x = a/e$ is the directrix corresponding to the focus Q.

How come though, the two directrices are the two lines $x = -a/e$, and $x = a/e$?

To begin with, in an ellipse, the distance from a point to a focus is less than the distance from the point to the directrix corresponding to the focus, since: **$0 < e < 1$ for an ellipse**.

However, in a hyperbola, the distance from a point to a focus is greater than the distance from the point to the directrix corresponding to the focus, because: **$e > 1$ for a hyperbola**.

So next, we can use the fact above the way below:

Suppose **d** is the distance from a point in a hyperbola to a focus, and **D** is the distance from the point to the directrix corresponding to the focus.

Then, we get: **$d = eD$** where e is the eccentricity. In other words, we get: $\dfrac{d}{D} = e$.

And the same is true for the other directrix, too. What then, are the two directrices?

Assuming **H** is the hyperbola below, we can say that the directrices of **H** are two vertical lines in gray, one is **D_L**, which is: **$x = -a/e$**, and the other is **D_R**, which is: **$x = a/e$**.

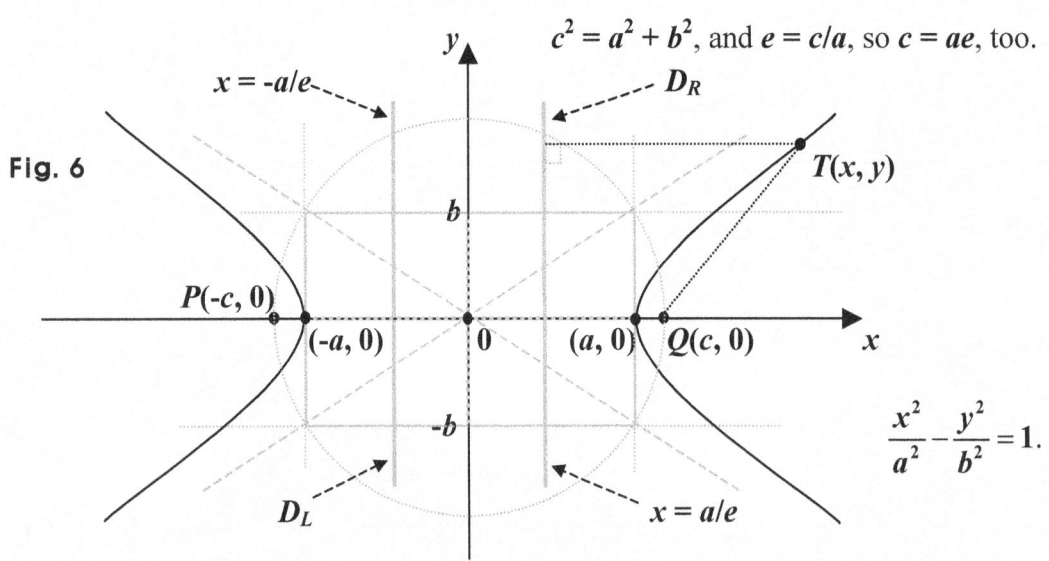

Fig. 6

Let's now check to see if the directrices of **H** are: **$x = -a/e$**, and **$x = a/e$**.

To begin with, we know the fact that **$d = eD$**, where e is the eccentricity, **d** is the distance from a point in a hyperbola to a focus, and **D** is the distance from the point to the directrix corresponding to the focus. So we are going to use the fact above.

Suppose first, $x = m$ is the equation of the directrix D_R shown in the figure above, and next, the arbitrary point $T(x, y)$ is now in the right branch.

Then first, we can see that $(x - m)$ is the distance from $T(x, y)$ to the directrix D_R.

And next, assuming q is the distance from $T(x, y)$ to the focus $Q(c, 0)$, we get:

$$q^2 = (x - c)^2 + (y - 0)^2 = (x - c)^2 + y^2 \Rightarrow q^2 = (x - c)^2 + y^2.$$

And we have: $\dfrac{x^2}{a^2} - \dfrac{y^2}{b^2} = 1$, and $c^2 = b^2 + a^2$.

So we get: $\dfrac{x^2}{a^2} - \dfrac{y^2}{b^2} = \dfrac{x^2}{a^2} - \dfrac{y^2}{c^2 - a^2} = \dfrac{x^2}{a^2} + \dfrac{y^2}{a^2 - c^2} = 1 \Rightarrow \dfrac{y^2}{a^2 - c^2} = 1 - \dfrac{x^2}{a^2} = \dfrac{a^2 - x^2}{a^2}$

$\Rightarrow y^2 = \dfrac{(a^2 - c^2)(a^2 - x^2)}{a^2}$. Thus now, getting back to this: $q^2 = (x - c)^2 + y^2$, we get:

$$\Rightarrow q^2 = (x - c)^2 + \dfrac{(a^2 - c^2)(a^2 - x^2)}{a^2} = \dfrac{a^2(x - c)^2 + (a^2 - c^2)(a^2 - x^2)}{a^2}.$$

Meanwhile, $a^2(x - c)^2 + (a^2 - c^2)(a^2 - x^2) = a^2(x^2 - 2cx + c^2) + a^4 - a^2x^2 - a^2c^2 + c^2x^2$

$= a^2x^2 - 2a^2cx + a^2c^2 + a^4 - a^2x^2 - a^2c^2 + c^2x^2 = (a^4 - 2a^2cx + c^2x^2) = (a^2 - cx)^2$.

So we get: $q^2 = \dfrac{(a^2 - cx)^2}{a^2} = \left(\dfrac{a^2 - cx}{a}\right)^2 = \left(a - \dfrac{cx}{a}\right)^2$.

And we know q is a distance.   So we get: $q = \left| a - \dfrac{cx}{a} \right|$.

And we know $T(x, y)$ is now in the right branch.   So we now have: $x \geq a$, and $c > a$.

Thus, we get: $x \geq a \Rightarrow \dfrac{x}{a} \geq 1 \Rightarrow \dfrac{cx}{a} \geq c \Rightarrow -\dfrac{cx}{a} \leq -c \Rightarrow a - \dfrac{cx}{a} \leq a - c < 0$, since $c > a$.

So we get: $a - \dfrac{cx}{a} < 0 \Rightarrow \dfrac{cx}{a} - a > 0$. And we have: $q = \left| a - \dfrac{cx}{a} \right|$.

Thus, we get: $q = \dfrac{cx}{a} - a.$ And we have: $e = c/a.$ So we get: $c = ae$, too.

Thus, we get: $q = \dfrac{cx}{a} - a = \dfrac{aex}{a} - a = ex - a \Rightarrow q = ex - a.$

Next, we have: $d = eD$, where d is the distance from a point in a hyperbola to a focus, and D is the distance from the point to the directrix corresponding to the focus.

And in this case, d is the distance from the arbitrary point $T(x, y)$ to the focus $Q(c, 0)$, and thus, is q, which is: $ex - a.$ So the distance d is: $ex - a.$

And D is the distance from the point $T(x, y)$ to the right directrix corresponding to the focus Q, and the right directrix is D_R, which is the line $x = m$. So the distance D is: $x - m.$

Thus, in sum, we have: $d = ex - a$, and $D = x - m.$ So we get:

$d = eD \Rightarrow ex - a = e(x - m) = ex - em \Rightarrow ex - a = ex - em \Rightarrow a = em \Rightarrow m = a/e.$

And the directrix is D_R, which is the line $x = m$.
So the directrix D_R is the line $x = a/e$.

Suppose this time, $x = -n$ is the equation of the directrix D_L as shown in the figure below, and the arbitrary point $T(x, y)$ is now in the left branch.

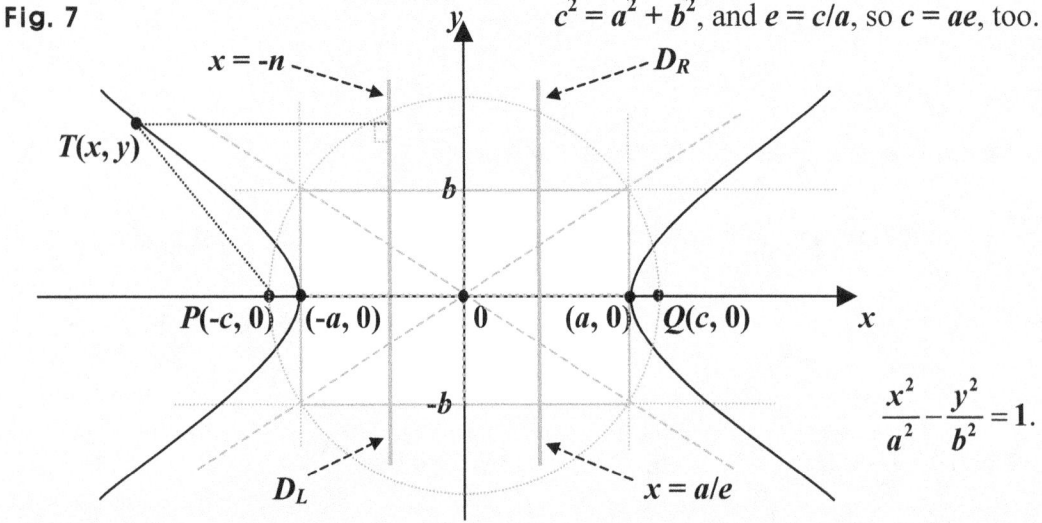

Fig. 7

$c^2 = a^2 + b^2$, and $e = c/a$, so $c = ae$, too.

$\dfrac{x^2}{a^2} - \dfrac{y^2}{b^2} = 1.$

Then first, we can see that $(-n - x)$ is the distance from $T(x, y)$ to the directrix D_L.

And next, assuming p is the distance from $T(x, y)$ to the focus $P(-c, 0)$, we get:

$$p^2 = (x + c)^2 + (y - 0)^2 = (x + c)^2 + y^2 \Rightarrow p^2 = (x + c)^2 + y^2.$$

And we know that since we have: $\dfrac{x^2}{a^2} - \dfrac{y^2}{b^2} = 1$, and $c^2 = b^2 + a^2$, we can get:

$$y^2 = \frac{(a^2 - c^2)(a^2 - x^2)}{a^2}.$$

So we get: $p^2 = (x + c)^2 + \dfrac{(a^2 - c^2)(a^2 - x^2)}{a^2} = \dfrac{a^2(x + c)^2 + (a^2 - c^2)(a^2 - x^2)}{a^2}.$

Meanwhile, $a^2(x + c)^2 + (a^2 - c^2)(a^2 - x^2) = a^2(x^2 + 2cx + c^2) + a^4 - a^2x^2 - a^2c^2 + c^2x^2$

$= a^2x^2 + 2a^2cx + a^2c^2 + a^4 - a^2x^2 - a^2c^2 + c^2x^2 = (a^4 + 2a^2cx + c^2x^2) = (a^2 + cx)^2.$

So we get: $p^2 = \dfrac{(a^2 + cx)^2}{a^2} = \left(\dfrac{a^2 + cx}{a}\right)^2 = \left(a + \dfrac{cx}{a}\right)^2.$

And we know p is a distance.   So we get: $p = \left| a + \dfrac{cx}{a} \right|.$

And we know $T(x, y)$ is now in the left branch.   So we now have: $x \leq -a$, and $c > a$.

Thus, we get: $x \leq -a \Rightarrow \dfrac{x}{a} \leq -1 \Rightarrow \dfrac{cx}{a} \leq -c \Rightarrow a + \dfrac{cx}{a} \leq a - c < 0$, since $c > a$.

So we get: $a + \dfrac{cx}{a} < 0 \Rightarrow -a - \dfrac{cx}{a} > 0$. And we have: $p = \left| a + \dfrac{cx}{a} \right|.$

Thus, we get: $p = -a - \dfrac{cx}{a}.$ And we have: $e = c/a$.   So we get: $c = ae$, too.

Thus, we get: $p = -a - \dfrac{cx}{a} = -a - \dfrac{aex}{a} = -a - ex \Rightarrow p = -(a + ex).$

Next, we have: $d = eD$, where d is the distance from a point in a hyperbola to a focus, and D is the distance from the point to the directrix corresponding to the focus.

And in this case, d is the distance from the arbitrary point $T(x, y)$ to the focus $P(-c, 0)$, and thus, is p, which is: $-(a + ex)$.   So the distance d is: $-(a + ex)$.

And D is the distance from the point $T(x, y)$ to the left directrix corresponding to the focus P, and the left directrix is D_L, which is the line $x = -n$. So the distance D is: $-n - x$.

Thus, in sum, we have: $d = -(a + ex)$, and $D = -(n + x)$.   So we get:

$d = eD \Rightarrow -(a + ex) = -e(n + x) \Rightarrow a + ex = e(n + x) = en + ex \Rightarrow a = en \Rightarrow n = a/e$.

And the left directrix is D_L, which is the line $x = -n$.

So the directrix D_L is the line $x = -a/e$.

And let's now find for instance, the directrices of this hyperbola: $\dfrac{x^2}{4^2} - \dfrac{y^2}{2^2} = 1$.

We have: $x = \pm a/e$, $e = c/a$, and $c = \sqrt{a^2 + b^2}$. So first, we can get: $x = \pm a^2/c$.
And we have: $a = 4$, and $b = 2$.

So finding first, the focal distance, we get: $c = \sqrt{a^2 + b^2} = \sqrt{4^2 + 2^2} = \sqrt{20} = 2\sqrt{5}$.

Thus, next, we can get: $a^2/c = \dfrac{16}{2\sqrt{5}} = \dfrac{8}{\sqrt{5}} = \dfrac{8\sqrt{5}}{5}$.

So the directrices are: $x = \pm \dfrac{8\sqrt{5}}{5} \approx \pm 3.5777$.

What then, about the directrices of this hyperbola: $\dfrac{(x - u)^2}{a^2} - \dfrac{(y - v)^2}{b^2} = 1$?

We know that the hyperbola H is: $\dfrac{x^2}{a^2} - \dfrac{y^2}{b^2} = 1$.

So assuming **G** is the hyperbola in question, we can get **G** translating the hyperbola **H** in the amount of **u** in the direction of the **x**-axis, and in the amount of *v* along the **y**-axis.

And we know that the center of the hyperbola **G** is (**u, v**).
So in short, moving the center of the hyperbola **H** to the point (**u, v**), we can get **G**.

Then, the directrices of **H** will be translated exactly the way the hyperbola **H** gets translated.

And we know that the directrices of **H** are: $x = \pm a/e$.

So the directrices of **G** will be as follows: $x = \pm a/e + u$.

That is, the two directrices are: $x = -a/e + u$, and $x = a/e + u$.

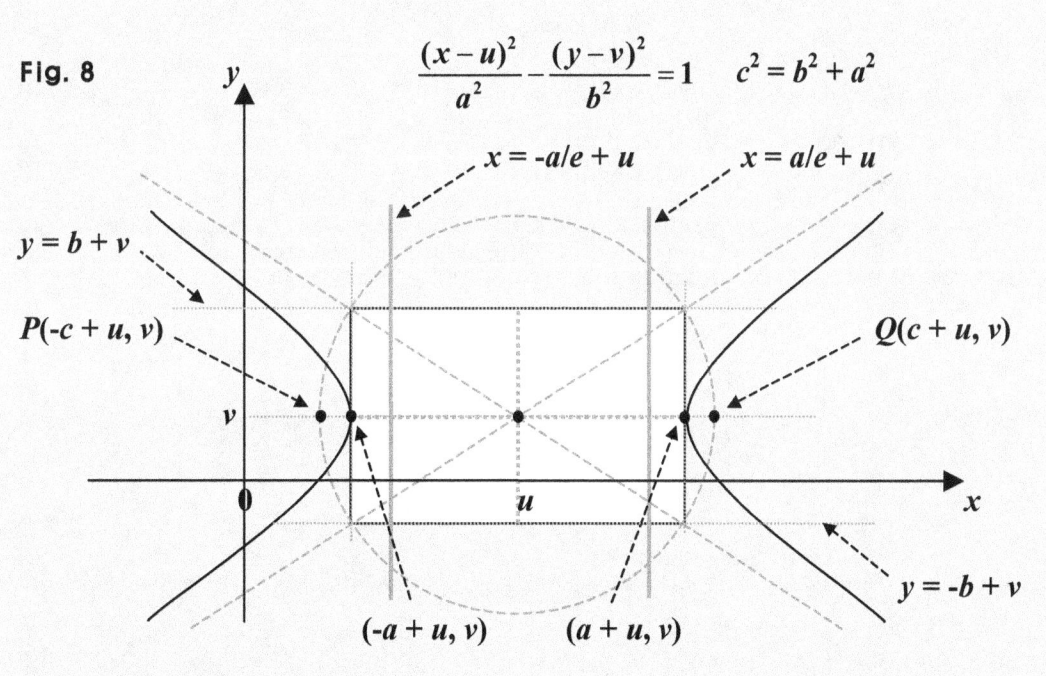

Fig. 8

$$\frac{(x-u)^2}{a^2} - \frac{(y-v)^2}{b^2} = 1 \qquad c^2 = b^2 + a^2$$

Let's now, for instance, find the directrices of this hyperbola: $\dfrac{(x-2)^2}{4^2} - \dfrac{(y-1)^2}{2^2} = 1$.

First, we've already found the directrices of $\dfrac{x^2}{2^2} - \dfrac{y^2}{4^2} = 1$, and they are: $x = \pm \dfrac{8\sqrt{5}}{5}$.

So next, translating the directrices above by 2 along the x-axis, and by 1 along the y-axis, we get the directrices of the hyperbola we want to find the directrices of.

We know however, translating a vertical line along the y-axis, we see no change in the line. So assuming $(2, 1)$ is a new center, we just add the x-coordinate of the new center to the directrices of the hyperbola centered at the origin.

Then, we get: $x = \pm \dfrac{8\sqrt{5}}{5} + 2$, which are the directrices we want. And more specifically:

One is: $x = -\dfrac{8\sqrt{5}}{5} + 2 = \dfrac{10 - 8\sqrt{5}}{5} \approx -1.58$, which corresponds to the left focus.

And the other is: $x = \dfrac{8\sqrt{5}}{5} + 2 = \dfrac{10 + 8\sqrt{5}}{5} \approx 5.58$, which corresponds to the right focus.

- What if the hyperbola is vertical, that is, if it is: $\dfrac{y^2}{b^2} - \dfrac{x^2}{a^2} = 1$?

Putting first, the hyperbola above in a graph, we can put it the way below:

Fig. 9

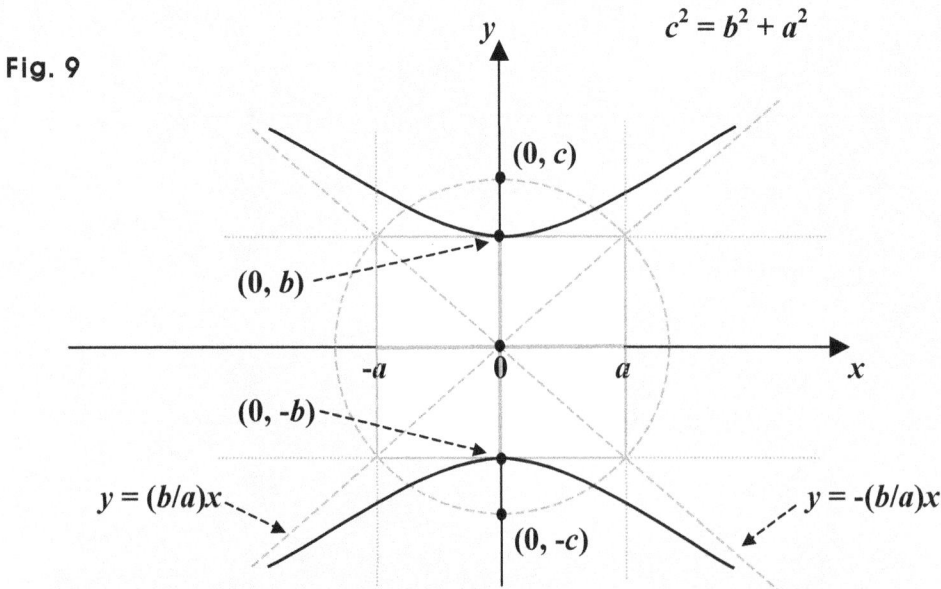

To begin with, we know the fact that if in a hyperbola, d is the distance from a point to a focus, and D is the distance from the point to the directrix corresponding to the focus, we get: $d = eD$, where e is the eccentricity.   In other words, we get: $\frac{d}{D} = e$.

Next, the directrices are perpendicular to the transverse axis.

And we know if a hyperbola is horizontal, the transverse axis is parallel to the x-axis. So the directrices of a hyperbola horizontal are perpendicular to the x-axis. What then, about the directrices of a hyperbola vertical?

If the hyperbola is vertical, the transverse axis is parallel to the y-axis. So the directrices are horizontal, that is, perpendicular to the y-axis.

And next, the directrices of H are: $x = \pm a/e$, where a is half the transverse axis, and e is the eccentricity. What then, are the directrices of the hyperbola vertical?

If the vertical hyperbola is K, the directrices of K are: $y = \pm b/e$, where b is half the transverse axis.

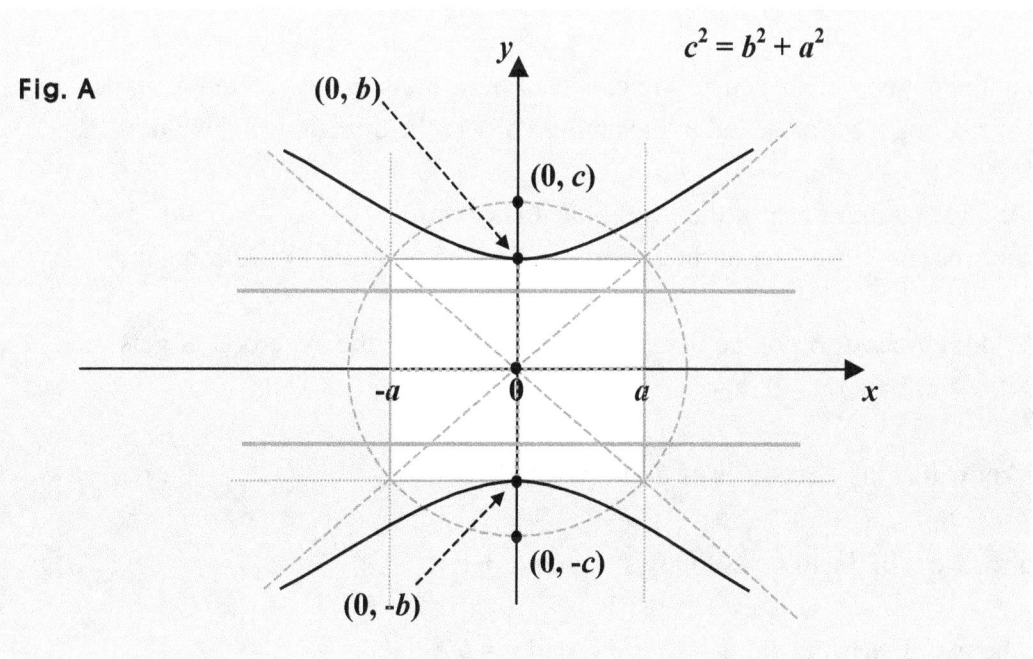

So for instance, what are the directrices of this hyperbola: $\dfrac{y^2}{4^2} - \dfrac{x^2}{2^2} = 1$?

We know finding the directrices of $\dfrac{x^2}{4^2} - \dfrac{y^2}{2^2} = 1$, we get: $x = \pm\dfrac{8\sqrt{5}}{5}$.

So the directrices of $\dfrac{y^2}{4^2} - \dfrac{x^2}{2^2} = 1$ are: $y = \pm\dfrac{8\sqrt{5}}{5}$. How come?

We have: $y = \pm b/e$, $e = c/b$, and $c = \sqrt{b^2 + a^2}$. So first, we can get: $y = \pm b^2/c$.

And we have: $a = 2$, and $b = 4$.

So finding first, the focal distance, we get: $c = \sqrt{a^2 + b^2} = \sqrt{4^2 + 2^2} = \sqrt{20} = 2\sqrt{5}$.

Thus, the directrices are: $y = \pm b^2/c = \pm\dfrac{16}{2\sqrt{5}} = \pm\dfrac{8}{\sqrt{5}} = \pm\dfrac{8\sqrt{5}}{5}$. That is, $y = \pm\dfrac{8\sqrt{5}}{5}$.

• What then, about the directrices of this hyperbola: $\dfrac{(y-v)^2}{b^2} - \dfrac{(x-u)^2}{a^2} = 1$?

Assuming the hyperbola above is *J*, we can get *J* translating the hyperbola *K* in the amount of *u* along the *x*-axis, and in the amount of *v* in the direction of the *y*-axis.

And we know that the center of the hyperbola *J* is (*u*, *v*).
So in short, moving the center of the hyperbola *K* to the point (*u*, *v*), we can get *J*.

Then, the directrices of *K* will be translated exactly the way the hyperbola *K* gets translated.

And we know that the directrices of *K* are: $y = \pm b/e$.

So the directrices of *J* will be as follows: $y = \pm b/e + v$.

That is, the two directrices are: $y = -b/e + v$, and $y = b/e + v$.

Let's now, for instance, find the directrices of this hyperbola: $\dfrac{(y+5)^2}{4^2} - \dfrac{(x-3)^2}{2^2} = 1$.

First, we've already found the directrices of $\dfrac{y^2}{4^2} - \dfrac{x^2}{2^2} = 1$, and they are: $y = \pm\dfrac{8\sqrt{5}}{5}$.

So next, translating those directrices by 3 along the x-axis, and by -5 along the y-axis, we get the directrices of the hyperbola we want to find the directrices of.

We know however, the directrices above are horizontal, and translating horizontal lines along the x-axis, we see no change in the lines.

So assuming (3, -5) is a new center, we just add the y-coordinate of the new center to the directrices above.
Note however, we can do so, because the hyperbola getting translated is centered at the origin.

Then, we get: $y = \pm\dfrac{8\sqrt{5}}{5} - 5$, which are the directrices we want. And more specifically:

One is: $y = -\dfrac{8\sqrt{5}}{5} - 5 \approx -8.58$, which corresponds to the lower focus.

And the other is: $y = \dfrac{8\sqrt{5}}{5} - 5 \approx -2.58$, which corresponds to the upper focus.

And translating the curve of $\dfrac{y^2}{b^2} - \dfrac{x^2}{a^2} = 1$ in the amount of u along the x-axis, and in

the amount of v along the y-axis, we get the curve of $\dfrac{(y-v)^2}{b^2} - \dfrac{(x-u)^2}{a^2} = 1$.

Besides, the equations of the asymptotes are as follows:

$y - v = \pm(b/a)(x - u)$, which are the two equations of the two asymptotes we get if we translate the asymptotes $y = \pm(b/a)x$ by u along the x-axis and by v along the y-axis.

And putting in graphs the two curves above along with their elements, we can put them the way below:

Fig. B

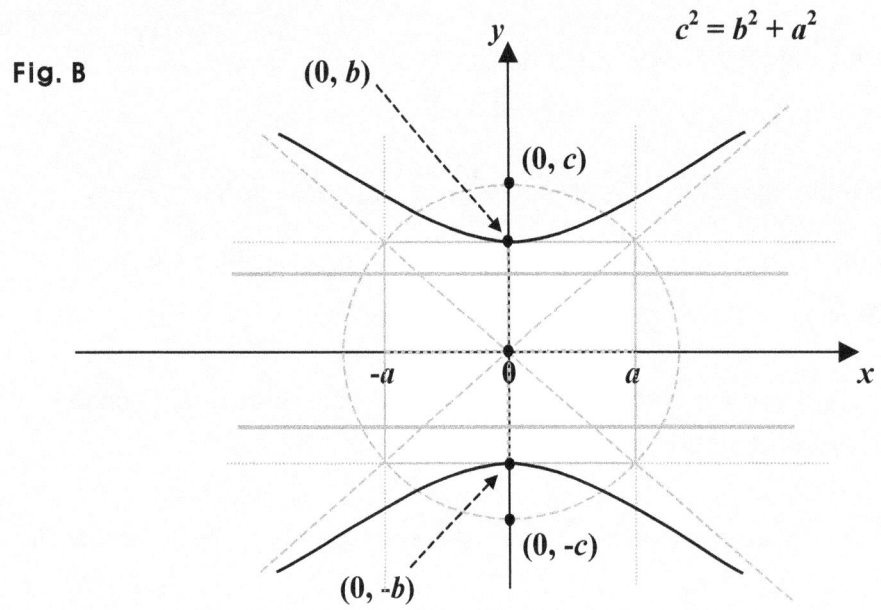

Fig. C

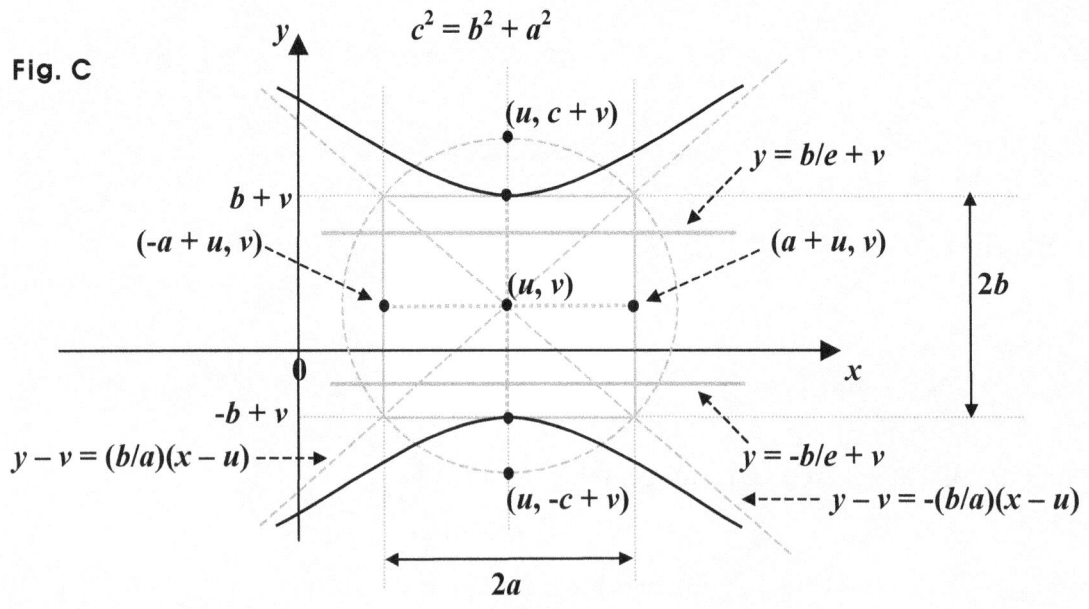

Examples 5 in Standard Forms

Find the center and the two main axes of each hyperbola below:

0. $x^2 + 2x - 2y^2 + 2y - 1 = 0$

1. $36y^2 + 36y - 100x^2 - 300x + 9 = 0$

2. $x^2 + 12x - 9y^2 - 90y + 252 = 0$

3. $2x - 2x^2 + 3y + y^2 + 1 = 0$

4. $25x^2 + 250x - 16y^2 + 16y + 521 = 0$

5. $9y^2 - 4x^2 + 24x = 0$

Suggestions or Solutions
To the Problems in the Examples

0. $x^2 + 2x - 2y^2 + 2y - 1 = 0 \Rightarrow x^2 + 2x + 1 - 2(y^2 - y) - 1 = 0$

$\Rightarrow (x+1)^2 - 2\{y^2 - y + (\frac{1}{2})^2 - (\frac{1}{2})^2\} - 1 = 0 \Rightarrow (x+1)^2 - 2\{y^2 - y + (\frac{1}{2})^2\} - \frac{1}{2} - 1 = 0$

$\Rightarrow (x+1)^2 - 2(y - \frac{1}{2})^2 - \frac{3}{2} = 0 \Rightarrow (x+1)^2 - 2(y - \frac{1}{2})^2 = \frac{3}{2} \Rightarrow \dfrac{(x+1)^2}{\frac{3}{2}} - \dfrac{2(y - \frac{1}{2})^2}{\frac{3}{2}} = 1$

$\Rightarrow \dfrac{(x+1)^2}{(\sqrt{\frac{3}{2}})^2} - \dfrac{(y - \frac{1}{2})^2}{\frac{3}{4}} = 1 \Rightarrow \dfrac{(x+1)^2}{(\sqrt{\frac{3}{2}})^2} - \dfrac{(y - \frac{1}{2})^2}{(\frac{\sqrt{3}}{2})^2} = 1.$

So the hyperbola is horizontal, the center is $(-1, \frac{1}{2})$, the transverse axis is $2\sqrt{\frac{3}{2}}$, which is $\sqrt{6}$, and the conjugate axis is $\sqrt{3}$.

1. $36y^2 + 36y - 100x^2 - 300x + 9 = 0 \Rightarrow 36(y^2 + y) - 100(x^2 + 3x) + 9 = 0$

$\Rightarrow 36(y^2 + y + \frac{1}{4} - \frac{1}{4}) - 100(x^2 + 3y + \frac{9}{4} - \frac{9}{4}) + 9 = 0$

$\Rightarrow 36(y^2 + y + \frac{1}{4}) - 9 - 100(x^2 + 3x + \frac{9}{4}) - 225 + 9 = 0$

$\Rightarrow 36(y + \frac{1}{2})^2 - 100(x + \frac{3}{2})^2 - 225 = 0 \Rightarrow 36(y + \frac{1}{2})^2 - 100(x + \frac{3}{2})^2 = 225$

$\Rightarrow \dfrac{36(y + \frac{1}{2})^2}{225} - \dfrac{100(x + \frac{3}{2})^2}{225} = 1 \Rightarrow \dfrac{4(y + \frac{1}{2})^2}{25} - \dfrac{4(x + \frac{3}{2})^2}{9} = 1 \Rightarrow \dfrac{(y + \frac{1}{2})^2}{\frac{25}{4}} - \dfrac{(x + \frac{3}{2})^2}{\frac{9}{4}} = 1$

$\Rightarrow \dfrac{(y + \frac{1}{2})^2}{(\frac{5}{2})^2} - \dfrac{(x + \frac{3}{2})^2}{(\frac{3}{2})^2} = 1.$

So the hyperbola is vertical, the center is $(-\frac{3}{2}, -\frac{1}{2})$, the transverse axis is 5, and the conjugate axis is 3.

2. $x^2 + 12x - 9y^2 - 90y + 252 = 0$

$\Rightarrow x^2 + 12x - 9y^2 - 90y + 252 = x^2 + 12x - 9(y^2 + 10y) + 252$

$= x^2 + 12x + 36 - 36 - 9(y^2 + 10y + 25 - 25) + 252$

$= (x + 6)^2 - 36 - 9(y^2 + 10y + 25) - 225 + 252$

$= (x + 6)^2 - 9(y + 5)^2 - 9 = 0$

$\Rightarrow \dfrac{(x+6)^2}{9} - (y+5)^2 = 1 \Rightarrow \dfrac{\{x-(-6)\}^2}{3^2} - \dfrac{\{y-(-5)\}^2}{1^2} = 1.$

So the hyperbola is horizontal, the center is (-6, -5), the transverse axis is 6, and the conjugate axis is 2.

3. $2x - 2x^2 + 3y + y^2 + 1 = 0 \Rightarrow -2x^2 + 2x + y^2 + 3y + 1 = 0$

$\Rightarrow -2x^2 + 2x + y^2 + 3y + 1 = -2(x^2 - x + \frac{1}{4} - \frac{1}{4}) + (y^2 + 3y + \frac{9}{4} - \frac{9}{4}) + 1$

$= -2(x - \frac{1}{2})^2 + \frac{1}{2} + (y + \frac{3}{2})^2 - \frac{9}{4} + 1 = -2(x - \frac{1}{2})^2 + (y + \frac{3}{2})^2 - \frac{3}{4} = 0$

$\Rightarrow -2(x - \frac{1}{2})^2 + (y + \frac{3}{2})^2 = \frac{3}{4} \Rightarrow \dfrac{(y+\frac{3}{2})^2}{\frac{3}{4}} - \dfrac{2(x-\frac{1}{2})^2}{\frac{3}{4}} = 1 \Rightarrow \dfrac{(y+\frac{3}{2})^2}{\frac{3}{4}} - \dfrac{(x-\frac{1}{2})^2}{\frac{3}{8}} = 1$

$\Rightarrow \dfrac{(y+\frac{3}{2})^2}{(\frac{\sqrt{3}}{2})^2} - \dfrac{(x-\frac{1}{2})^2}{(\frac{\sqrt{3}}{2\sqrt{2}})^2} = 1 \Rightarrow \dfrac{(y+\frac{3}{2})^2}{(\frac{\sqrt{3}}{2})^2} - \dfrac{(x-\frac{1}{2})^2}{(\frac{\sqrt{6}}{4})^2} = 1.$

So the hyperbola is vertical, the center is $(\frac{1}{2}, -\frac{3}{2})$, the transverse axis is $\sqrt{3}$, and the conjugate axis is $\frac{\sqrt{6}}{2}$.

4. $25x^2 + 250x - 16y^2 + 16y + 521 = 0$

$\Rightarrow 25x^2 + 250x - 16y^2 + 16y + 521 = 25(x^2 + 10x) - 16(y^2 - y) + 521$

$= 25(x^2 + 10x + 25 - 25) - 16(y^2 - y + \frac{1}{4} - \frac{1}{4}) + 521$

$= 25(x^2 + 10x + 25) - 625 - 16(y^2 - y + \frac{1}{4}) + 4 + 521$

$= 25(x+5)^2 - 16(y - \frac{1}{2})^2 - 100 = 0 \Rightarrow 25(x+5)^2 - 16(y - \frac{1}{2})^2 = 100$

$\Rightarrow \dfrac{25(x+5)^2}{100} - \dfrac{16(y-\frac{1}{2})^2}{100} = 1 \Rightarrow \dfrac{(x+5)^2}{4} - \dfrac{4(y-\frac{1}{2})^2}{25} = 1 \Rightarrow \dfrac{(x+5)^2}{4} - \dfrac{(y-\frac{1}{2})^2}{\frac{25}{4}} = 1$

$\Rightarrow \dfrac{\{x-(-5)\}^2}{2^2} - \dfrac{(y-\frac{1}{2})^2}{(\frac{5}{2})^2} = 1.$

So the hyperbola is horizontal, the center is $(-5, \frac{1}{2})$, the transverse axis is 4, and the conjugate axis is 5.

5. $9y^2 - 4x^2 + 24x - 72 = 0$

$\Rightarrow 9y^2 - 4x^2 + 24x - 72 = 9y^2 - 4(x^2 - 6x) - 72 = 9y^2 - 4(x^2 - 6x + 9 - 9) - 72$

$= 9y^2 - 4(x^2 - 6y + 9) + 36 - 72 = 9y^2 - 4(x-3)^2 - 36 = 0 \Rightarrow 9y^2 - 4(x-3)^2 = 36$

$\Rightarrow \dfrac{9y^2}{36} - \dfrac{4(x-3)^2}{36} = 1 \Rightarrow \dfrac{y^2}{4} - \dfrac{(x-3)^2}{9} = 1 \Rightarrow \dfrac{y^2}{2^2} - \dfrac{(x-3)^2}{3^2} = 1.$

So the hyperbola is vertical, the center is (3, 0), the transverse axis is 4, and the conjugate axis is 6.

₄.Summary on Hyperbolas

Now, summing up, we can put together the ideas on hyperbolas the way below:

To begin with, putting a hyperbola in a standard equation, we can get: $\dfrac{x^2}{a^2} - \dfrac{y^2}{b^2} = 1$.

What hyperbola then, is it?

If **H** is the hyperbola, **H** is <u>centered at (0, 0)</u>, the origin, and since the coefficient of x^2 is positive, it is <u>horizontal</u>, so the transverse axis is parallel to the **x**-axis, and is **2a**, and the conjugate axis is **2b**.   And assuming **c** is the <u>focal distance</u>, we get: $c^2 = b^2 + a^2$.

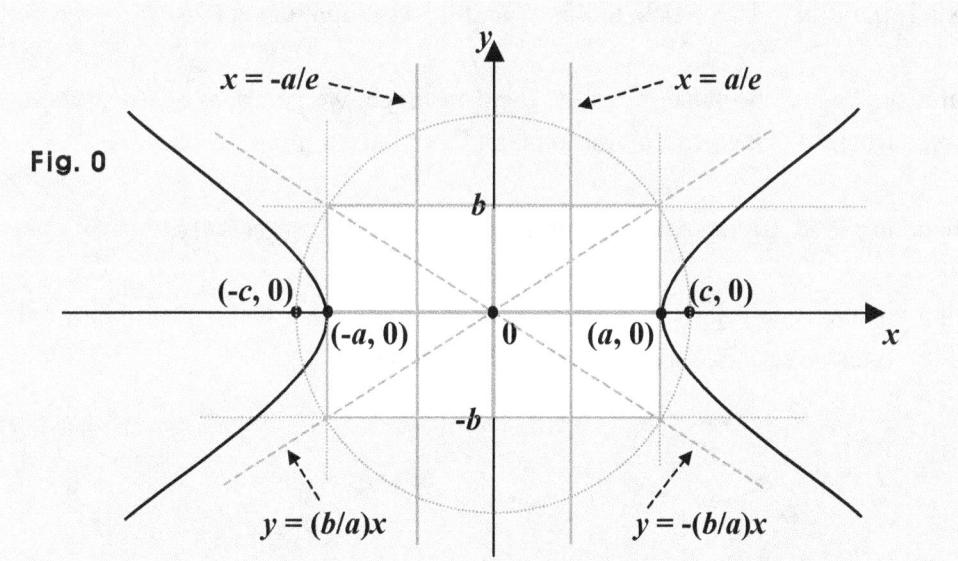

Fig. 0

So next, since the semi major axis is **a**, assuming **e** is the <u>eccentricity</u>, we get: $e = c/a$, where **c** is the focal distance, of course.   And we have: $e > 1$, since $c > a$.

And next, since **H** is horizontal, the two directrices are two lines parallel to the **y**-axis. The distance from the center to each directrix is: **a/e**, where **a** is <u>half the transverse axis</u>, and **e** is the eccentricity. So since the center is **(0, 0)**, the two directrices are: $x = \pm a/e$. And we know: $e = c/a$. So we can put the two directrices this way, too: $x = \pm a^2/c$.

And since the center is **(0, 0)**, the two asymptotes are: $y = \pm(b/a)x$.

Next, since the hyperbola **H** is horizontal, the center and foci have the same *y*-coordinate. And the center is the midpoint between the foci. So since the center is **(0, 0)**, and the focal distance is *c*, the two <u>foci</u> are **(-c, 0)** and **(c, 0)**.

And also, since **H** is horizontal, the center and the vertices share the same *y*-coordinate, too. And the center is the midpoint between the vertices. The distance from the center to a vertex is the semi major axis, which is half the transverse axis. So since half the transverse axis is *a*, and the center is **(0, 0)**, the two <u>vertices</u> are **(-a, 0)** and **(a, 0)**.

So for instance, putting a horizontal hyperbola in an equation, we can get: $\dfrac{x^2}{4^2} - \dfrac{y^2}{3^2} = 1$.
What hyperbola then, is it?

If **K** is the hyperbola, **K** is <u>centered at (0, 0)</u>, the origin, and is <u>horizontal</u>, so the transverse axis is parallel to the *x*-axis, and is 8, and the conjugate axis is 6.

And the focal distance is 5, because if *c* is the <u>focal distance</u>, we get: $c^2 = b^2 + a^2$, where **a = 4**, and **b = 3**, where *a* is the semi major axis, and *b* is the semi minor axis.

And since the center is **(0, 0)**, the two asymptotes are $y = \pm(b/a)x = \pm(3/4)x$.

Next, assuming *e* is the <u>eccentricity</u>, we get: $e = c/a = 5/4$, where *a* is the semi major axis, which is half the transverse axis, and *c* is the focal distance.

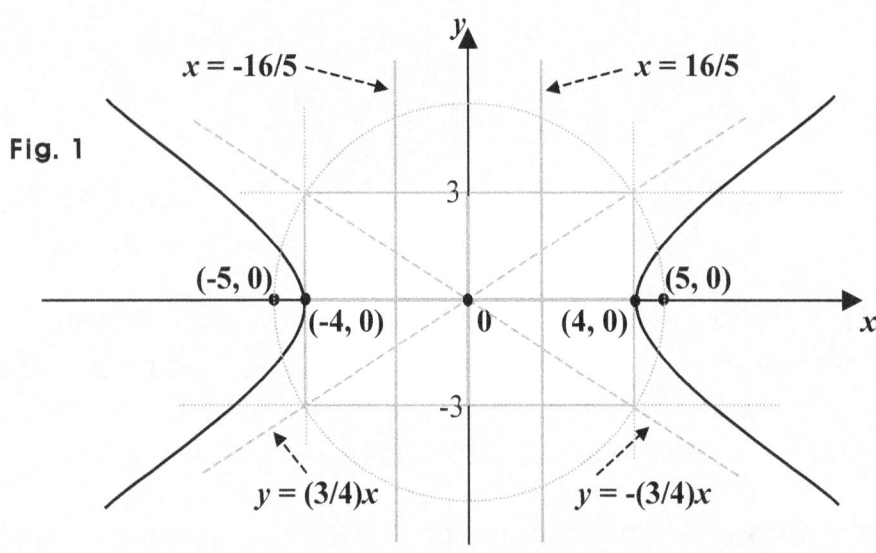

Fig. 1

And since the center is **(0, 0)**, the two directrices are: $x = \pm 16/5$.

Next, since the hyperbola **K** is horizontal, the center is (0, 0), and the focal distance is 5, the two <u>foci</u> are (-5, 0) and (5, 0).   (Note that horizontal means the same y-coordinate.)

And also, since **K** is horizontal, half the transverse axis, that is, the semi major axis is 4, and the center is (0, 0), the two <u>vertices</u> are (-4, 0) and (4, 0).

What then, about the hyperbola as follows: $\dfrac{(x-u)^2}{a^2} - \dfrac{(y-v)^2}{b^2} = 1$?

If **G** is the hyperbola, the hyperbola **G** is <u>centered at (u, v)</u>, and since the coefficient of x^2 is positive, **G** is <u>horizontal</u>, so the transverse axis is parallel to the x-axis, and is **2a**, and the conjugate axis is parallel to the y-axis, and is **2b**.

And assuming c is the <u>focal distance</u>, we get: $c^2 = b^2 + a^2$, where **a** is half the transverse axis, and **b** is half the conjugate axis.

So next, since half the transverse axis is **a**, assuming e is the <u>eccentricity</u>, we get: $e = c/a$, where **c** is the focal distance, of course.   And also, we have: **e > 1**, too, since **c > a**.

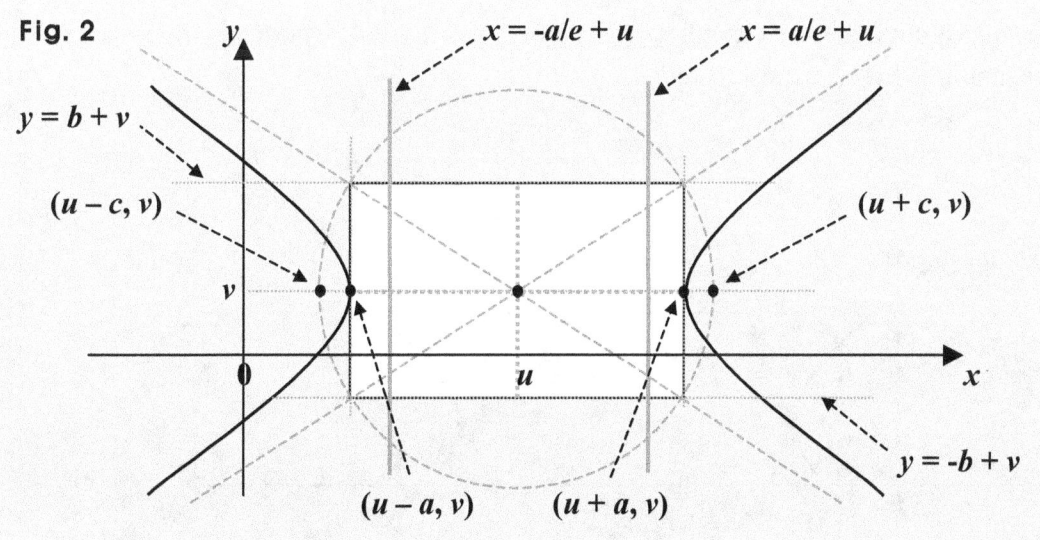

Fig. 2

And since the center is **(u, v)**, the two asymptotes are: $y - v = \pm(b/a)(x - u)$.
And also, since the center is **(u, v)**, the two directrices are: $x = \pm a/e + u = \pm a^2/c + u$.

Next, since the hyperbola **G** is horizontal, the center is at **(u, v)**, and the focal distance is **c**, the two <u>foci</u> are **(u − c, v)** and **(u + c, v)**.

And also, since **G** is horizontal, the center is **(u, v)**, and the semi major axis, that is, half the transverse axis is **a**, the two <u>vertices</u> are **(u − a, v)** and **(u + a, v)**.

And notice that translating the hyperbola **H** in the amount of **u** along the *x*-axis, and in the amount of **v** along the *y*-axis, we get the hyperbola **G**.

And also, of course, translating the hyperbola **G** in the amount of **-u** along the *x*-axis, and in the amount of **-v** along the *y*-axis, we get the hyperbola **H**.

So for instance, putting a horizontal hyperbola in an equation, we can get:

$$\frac{(x-7)^2}{4^2} - \frac{(y-2)^2}{3^2} = 1.$$ What hyperbola then, is it?

If **L** is the hyperbola, the hyperbola **L** is <u>centered at (7, 2)</u>, and is <u>horizontal</u>, so the transverse axis is parallel to the *x*-axis, and is 8, and the conjugate axis is 6.

And assuming **c** is the focal distance, we get: $c^2 = b^2 + a^2$, where **a = 4**, and **b = 3**, where **a** is half the transverse axis, and **b** is half the conjugate axis.   So the <u>focal distance</u> is 5.

Thus next, assuming **e** is the <u>eccentricity</u>, we get: **e = c/a = 5/4**, where **c** is the focal distance, and **a** is half the transverse axis.

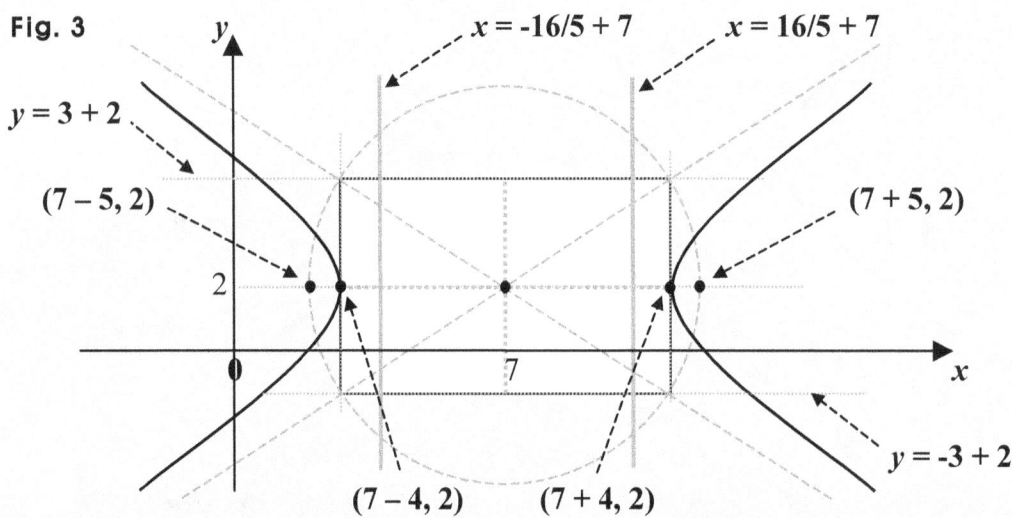

Fig. 3

$x = -16/5 + 7$ $x = 16/5 + 7$

$y = 3 + 2$

$(7 - 5, 2)$ $(7 + 5, 2)$

$(7 - 4, 2)$ $(7 + 4, 2)$

$y = -3 + 2$

And since the center is **(7, 2)**, the two asymptotes are: $y - 2 = \pm(3/4)(x - 7)$.

And also, since the center is **(7, 2)**, the two directrices are: $x = \pm16/5 + 7$.

Next, since the hyperbola **L** is horizontal, the focal distance is 5 and the center is (7, 2), the two <u>foci</u> are $(7 - 5, 2)$ and $(7 + 5, 2)$.

And also, since **L** is horizontal, the center is (7, 2), and the semi major axis, that is, half the transverse axis is 4, the two <u>vertices</u> are $(7 - 4, 2)$ and $(7 + 4, 2)$.

And next, putting a vertical hyperbola in a standard equation, we can get: $\dfrac{y^2}{b^2} - \dfrac{x^2}{a^2} = 1.$

What hyperbola then, is it?

If **V** is the hyperbola, **V** is <u>centered at **(0, 0)**</u>, the origin, and since the coefficient of y^2 is positive, it is <u>vertical</u>, so the transverse axis is parallel to the y-axis, and is **2b**, and the conjugate axis is **2a**.   And assuming **c** is the <u>focal distance</u>, we get: $c^2 = b^2 + a^2$.

Fig. 4

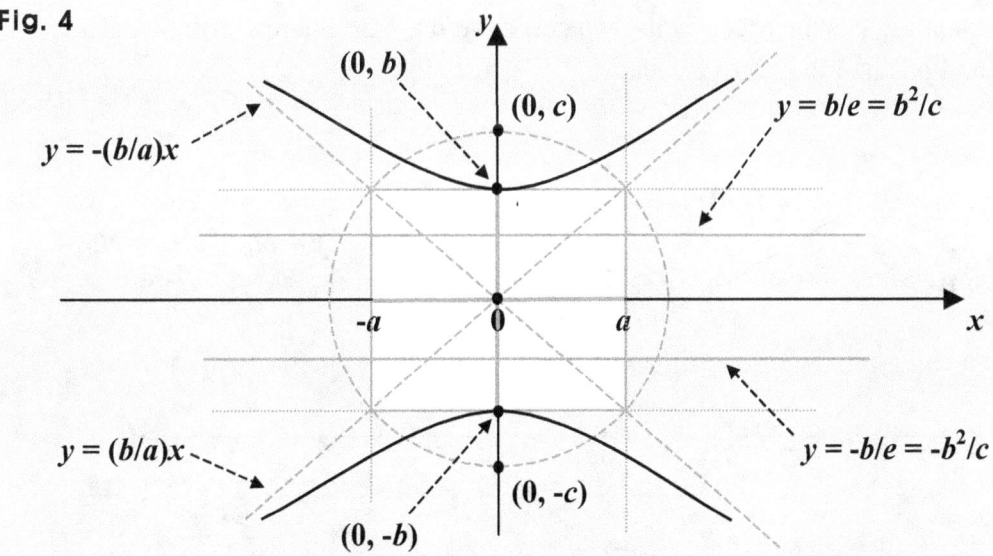

So next, assuming **e** is the eccentricity, we get: $e = c/b$, where **c** is the focal distance, and **b** is half the transverse axis.   And also, we have: $e > 1$, too, since $c > b$.

And since the center is **(0, 0)**, the two asymptotes are: $y = \pm(b/a)x$.

And also, since the center is **(0, 0)**, the two directrices are: $y = \pm b/e = \pm b^2/c$.

Next, since the hyperbola *V* is vertical, the center and foci share the same *x*-coordinate. So since the center is **(0, 0)**, and the focal distance is *c*, the two <u>foci</u> are **(0, *c*)** and **(0, -*c*)**.

And also, since *V* is vertical, the center and the vertices share the same *x*-coordinate, too. So since the semi major axis, that is, half the transverse axis is *b*, and the center is **(0, 0)**, the two <u>vertices</u> are **(0, *b*)** and **(0, -*b*)**.

So for instance, putting a vertical hyperbola in an equation, we can get: $\dfrac{y^2}{3^2} - \dfrac{x^2}{4^2} = 1$. What hyperbola then, is it?

If *M* is the hyperbola, *M* is <u>centered at (0, 0)</u>, the origin, and is <u>vertical</u>, so the transverse axis is parallel to the *y*-axis, and is 6, and the conjugate axis is 8.

And assuming *c* is the focal distance, we get: $c^2 = b^2 + a^2$, where *b* = **3**, and *a* = **4**, where *b* is half the transverse axis, and *a* is half the conjugate axis.   So the <u>focal distance</u> is 5.

So next, assuming *e* is the <u>eccentricity</u>, we get: *e* = *c*/*b* = **5/3**, where *c* is the focal distance, and *b* is half the transverse axis.

Fig. 5

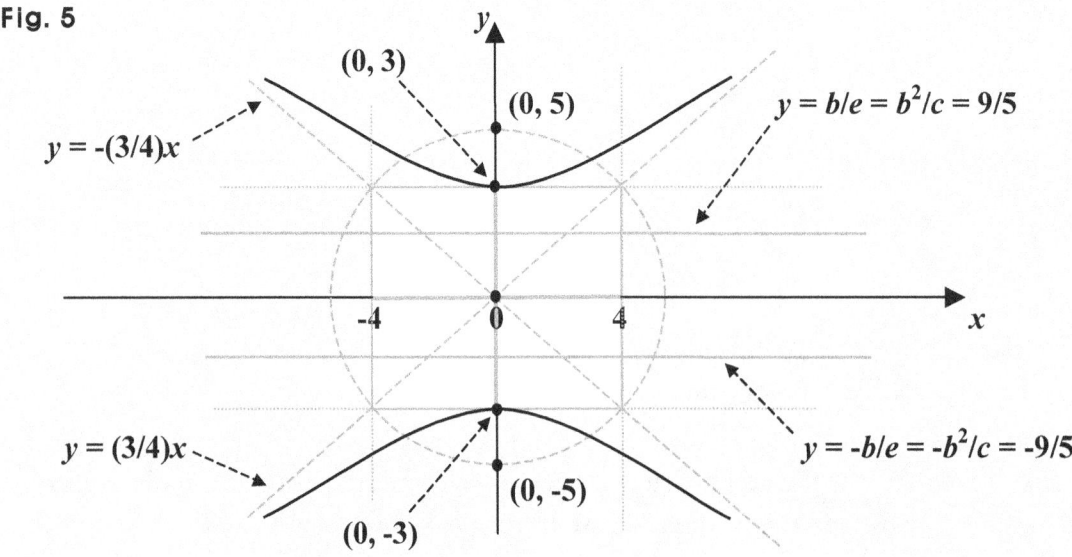

And since the center is **(0, 0)**, the two asymptotes are: $y = \pm(b/a)x = \pm(3/4)x$.
And also, since the center is **(0, 0)**, the two directrices are: $y = \pm b/e = \pm b^2/c = \pm 9/5$.

Next, since the hyperbola **M** is vertical, the center is (0, 0), and the focal distance is 5, the two <u>foci</u> are (0, 5) and (0, -5).   (Note that vertical means the same **x**-coordinate.)

And also, since **M** is vertical, the center is (0, 0), and the semi major axis, that is, half the transverse axis is 3 and, the two <u>vertices</u> are (0, 3) and (0, -3).

What then, about the hyperbola as follows: $\dfrac{(y-v)^2}{b^2} - \dfrac{(x-u)^2}{a^2} = 1$?

If **J** is the hyperbola, **J** is <u>centered at **(u, v)**</u>, and is <u>vertical</u>, so the transverse axis is parallel to the **y**-axis, and is **2b**, and the conjugate axis is parallel to the **x**-axis, and is **2a**.

And assuming **c** is the <u>focal distance</u>, we get: $c^2 = b^2 + a^2$.

So next, assuming **e** is the <u>eccentricity</u>, we get: **e = c/b**, where **c** is the focal distance, and half the transverse axis is **b**.   And also, we have: **e > 1**, too, since **c > b**.

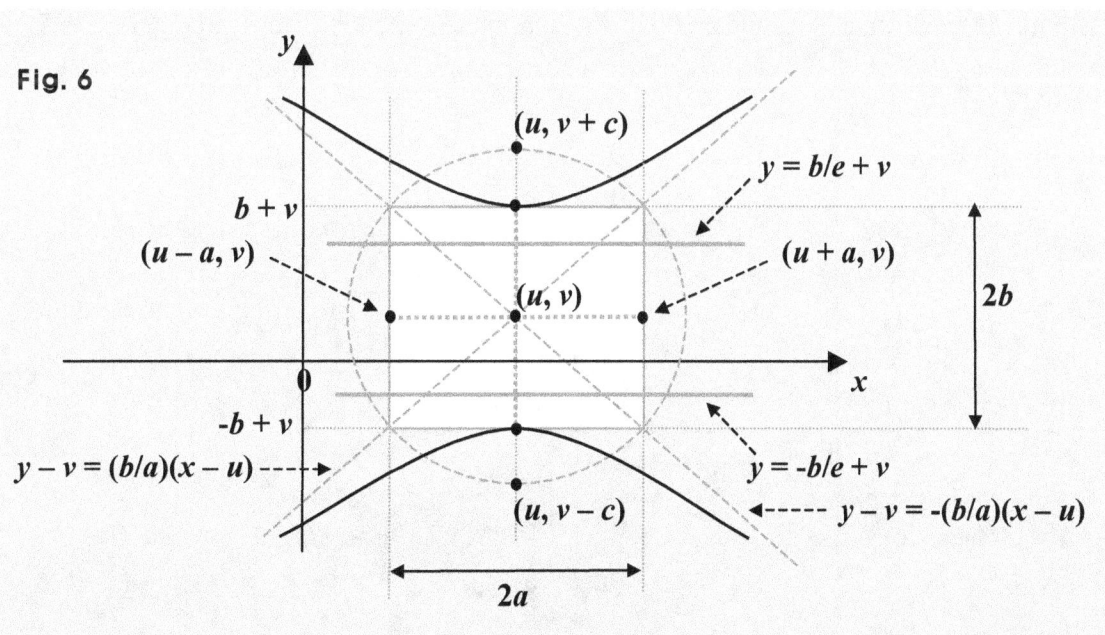

Fig. 6

And since the center is **(u, v)**, the two asymptotes are: $y - v = \pm(b/a)(x - u)$.
And also, since the center is **(u, v)**, the two directrices are: $y = \pm b/e + v = \pm b^2/c + v$.

Next, since *J* is vertical, the center and foci share the same *x*-coordinate. So since the center is (*u*, *v*), and the focal distance is *c*, the two <u>foci</u> are (*u*, *v* + *c*) and (*u*, *v* − *c*).

And also, since *J* is vertical, the center and the vertices share the same *x*-coordinate, too. So since the major radius, that is, the semi major axis is *b*, and the center is (*u*, *v*), the two <u>vertices</u> are (*u*, *v* + *b*) and (*u*, *v* − *b*).

So for instance, putting a vertical hyperbola in an equation, we can get:

$$\frac{(y-1)^2}{3^2} - \frac{(x-6)^2}{4^2} = 1.$$ What hyperbola then, is it?

If *N* is the hyperbola, *N* is <u>centered at (6, 1)</u>, and is <u>vertical</u>, so the transverse axis is parallel to the *y*-axis, and is 6, and the conjugate axis is 8.

And assuming *c* is the focal distance, we get: $c^2 = b^2 + a^2$, where *b* = 3, and *a* = 4, where *b* is half the transverse axis, and *a* is half the conjugate axis.   So the <u>focal distance</u> is 5.

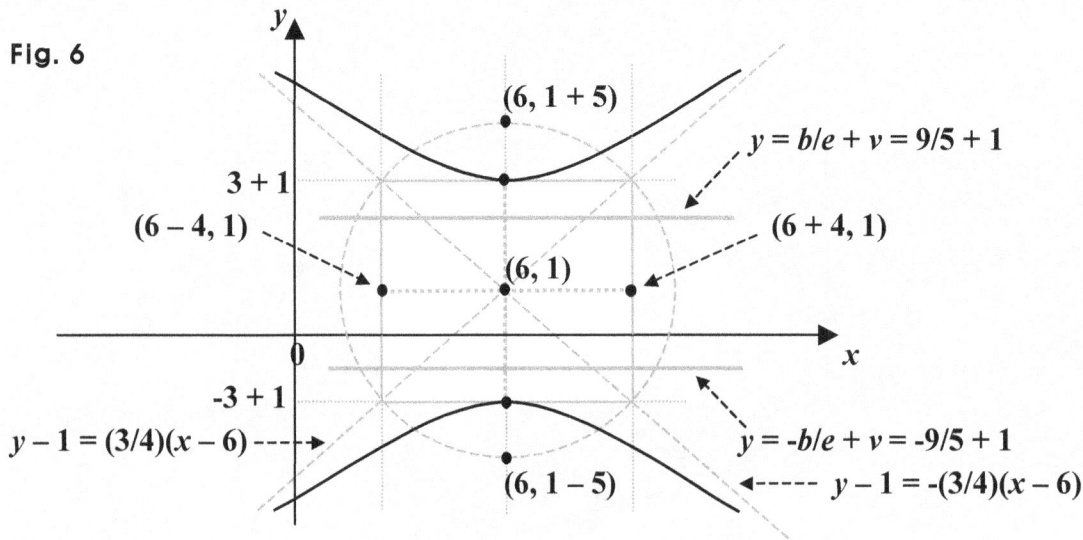

Fig. 6

And since the center is (6, 1), the two asymptotes are: *y* − 1 = ±(3/4)(*x* − 6).

And also, since the center is (6, 1), the two directrices are: *y* = ±*b*/*e* + *v* = ±9/5 + 1.

Next, since the hyperbola N is vertical, the center and foci share the same x-coordinate. So since the center is (5, 1), and the focal distance is 5, the two <u>foci</u> are (6, 6) and (6, -4).

And also, since N is vertical, the center and the vertices share the same x-coordinate, too. So since the center is (6, 1), and the semi major axis, that is, half the transverse axis is 3, the two <u>vertices</u> are (6, 3 + 1) and (6, -3 + 1).

And we call all the equations above <u>standard equations</u> for hyperbolas, and the hyperbolas above can be said to be perpendicular, because each main axis is perpendicular to a coordinate axis.

What about an equation then, that can indicate an hyperbola <u>perpendicular or not</u>?

Putting an hyperbola in such an equation, we can put it the way below:

$$ax^2 + by^2 + cxy + ux + vy + w = 0, \text{ where } 4ab < c^2.$$

Note that <u>a and b can be the same</u>. And of course, a, b, c, u, v, and w are constant.

Note <u>however, for some values of the constants, the equation does not indicate an hyperbola, even if $4ab < c^2$.</u>

For instance, $x^2 + 3y^2 + 4xy + 3x + 7y + 2 = 0$ is not an equation of a hyperbola but an equation of two lines. One is: $x + 3y + 1 = 0$, and the other is: $x + y + 2 = 0$.

That's because the expression $x^2 + 3y^2 + 4xy + 3x + 7y + 2$ can be factorized into the product of the two expressions $x + 3y + 1$ and $x + y + 2$.

That is to say that we can get: $x^2 + 3y^2 + 4xy + 3x + 7y + 2 = (x + 3y + 1)(x + y + 2) = 0$.

So we can get: $x + 3y + 1 = 0$, or $x + y + 2 = 0$.

And doing the factorization, we can take $x^2 + 3y^2 + 4xy + 3x + 7y + 2 = 0$ as a quadratic equation with respect to x or y, and then, apply the quadratic formula.

What then, about an equation that is in the <u>general form</u> and indicates an hyperbola <u>perpendicular only</u>?

Putting an hyperbola in such an equation, we can put it the way below:

$$ax^2 + by^2 + ux + vy + w = 0, \text{ where } ab < 0.$$

So the equation above indicates a hyperbola with a main axis perpendicular to the *x*-axis. Notice that it does not have the *xy*-term, that is, *cxy* in the equation shown earlier.

Note <u>however, for some values of the constants, the equation does not indicate a hyperbola, even if *ab* < 0</u>.

For instance, $x^2 - y^2 - 2x + 2y - 1 = 0$ and of course, $x^2 - 9y^2 - 1 = 0$ are hyperbola equations, but $x^2 - y^2 - 2x + 2y = 0$ and $x^2 - 9y^2 = 0$ are not.

That's because: $x^2 - 9y^2 = 0 \Rightarrow (x - 3y)(x + 3y) = 0$ which indicates not a hyperbola but two lines, one is: $x - 3y = 0$, and the other is: $x + 3y = 0$.

And next, moving on to the next equation, we get:

$$x^2 - y^2 - 2x + 2y = (x^2 - 2x) - (y^2 - 2y) = (x^2 - 2x + 1 - 1) - (y^2 - 2y + 1 - 1)$$

$$= (x - 1)^2 - (y - 1)^2 = 0 \Rightarrow \{x - 1 - (y - 1)\}\{x - 1 + (y - 1)\} = (x - y)(x + y - 2) = 0,$$
which indicates two lines, one is: $x - y = 0$, and the other is: $x + y - 2 = 0$.

So putting more specifically a perpendicular hyperbola in a general equation, we can put it the way below:

$$ax^2 + by^2 + ux + vy + w = 0, \text{ where } a > 0, b > 0, \text{ and } bu^2 + av^2 \neq 4abw.$$

That's simply because, putting the equation into the sum of perfect squares, we get:

$$ax^2 + by^2 + ux + vy + w = ax^2 + ux + by^2 + vy + w$$

$$= a(x^2 + \frac{u}{a}x) + b(y^2 + \frac{v}{b}y) + w = a(x^2 + \frac{u}{a}x + \frac{u^2}{4a^2} - \frac{u^2}{4a^2}) + b(y^2 + \frac{v}{b}y + \frac{v^2}{4b^2} - \frac{v^2}{4b^2}) + w$$

$$= a(x + \frac{u}{2a})^2 - \frac{u^2}{4a} + b(y + \frac{v}{2b})^2 - \frac{v^2}{4b} + w = 0$$

$$\Rightarrow a(x + \frac{u}{2a})^2 + b(y + \frac{v}{2b})^2 = \frac{u^2}{4a} + \frac{v^2}{4b} - w \neq 0.$$

Meanwhile: $\dfrac{u^2}{4a} + \dfrac{v^2}{4b} - w = \dfrac{bu^2 + av^2 - 4abw}{4ab}.$

So we get: $\dfrac{u^2}{4a} + \dfrac{v^2}{4b} - w \neq 0 \Rightarrow \dfrac{bu^2 + av^2 - 4abw}{4ab} \neq 0 \Rightarrow bu^2 + av^2 - 4abw \neq 0,$ for $ab \neq 0.$

Thus, we get: $bu^2 + av^2 \neq 4abw.$

Examples 1 in Hyperbolas

Assuming in each example below, C is the center of a hyperbola, F is a focus, V is a vertex, T is half the transverse axis, and t is half the conjugate axis, find the hyperbola, the eccentricity, the asymptotes, and the directrices, and put in a graph the hyperbola, together with the foci, vertices, asymptotes, and directrices.

0. $C(0, 0)$, $F(3, 0)$, and $V(2, 0)$.

1. $C(0, 0)$, $F(0, 3)$, and $V(0, 2)$.

2. $C(0, 0)$, $F(3, 0)$, and $t = 2$.

Suggestions or Solutions
To the Problem in the Example 0

Assuming $C(0, 0)$ is the center of a hyperbola, $F(3, 0)$ is a focus, and $V(2, 0)$ is a vertex, find the hyperbola, and its elements, and put them all in a graph.

To begin with, the center is $(0, 0)$, one focus is $(3, 0)$, and the y-coordinates are the same. So the hyperbola is horizontal, the other focus is $(-3, 0)$, the other vertex is $(-2, 0)$, and assuming c is the focal distance, we get: $c = 3$. Then, assuming a is half the transverse axis, and b is half the conjugate axis, we get:

$a = 2$, and $c^2 = a^2 + b^2 \Rightarrow 3^2 = 2^2 + b^2 \Rightarrow b^2 = 5$.

So the transverse axis is **4**, the conjugate axis is $2\sqrt{5}$, and the hyperbola is: $\dfrac{x^2}{4} - \dfrac{y^2}{5} = 1$.

And the asymptotes are: $y = \pm(b/a)x = \pm\frac{\sqrt{5}}{2}x$.

Next, assuming e is the eccentricity, we get: $e = c/a = 3/2$.

And next, the directrices are: $x = \pm a/e = \pm a^2/c = \pm 4/3$.

If not quite sure of the idea behind the processes above, follow the steps below:

To begin with, the hyperbola we want to find is centered at the origin. And if it is horizontal, the equation is: $\dfrac{x^2}{a^2} - \dfrac{y^2}{b^2} = 1$. If it is vertical, the equation is: $\dfrac{y^2}{b^2} - \dfrac{x^2}{a^2} = 1$.

So if finding if the hyperbola is horizontal or vertical, and the values of a and b, we find the hyperbola. How then, can we get them?

To begin with, if the hyperbola is horizontal, the coefficient of x^2 is positive. Then, a is half the transverse axis, and b is half the conjugate axis.

If it is vertical however, the coefficient of y^2 is positive. Then, a is half the conjugate axis, and b is half the transverse axis.

Next, the center is $(0, 0)$, and one of the foci is $(3, 0)$. So we can notice that the center and the foci share the same y-coordinate, which is 0. We can see thus, the hyperbola is horizontal.

So first, assuming the other focus is $(p, 0)$, since the center is $(0, 0)$, and is the midpoint between the foci, we get: $0 = (p + 3)/2$.   The other focus is thus, $(-3, 0)$.

Next, assuming c is the focal distance, we get: $c = 3$, because the focal distance is the distance from the center to a focus.

Next, if the hyperbola is <u>horizontal</u>, the center and vertices share the <u>same y-coordinate</u>, which is 0 in this case. And one vertex is $(2, 0)$.

So assuming the other vertex is $(q, 0)$, since the center is $(0, 0)$, and is the midpoint between the vertices, too, we get: $0 = (q + 2)/2$.   The other vertex is thus, $(-2, 0)$.

Next, we can say that **a is half the transverse axis**, which is the distance from a vertex to the center, and thus, is 2. So the transverse axis is 4.   What then, about **b**?

We have: $c^2 = a^2 + b^2$.   So we get: $3^2 = 2^2 + b^2 \Rightarrow b^2 = 9 - 4 = 5$.

And **$2b$** is the conjugate axis, which is thus, $2\sqrt{5}$.

So the hyperbola is: $\dfrac{x^2}{4} - \dfrac{y^2}{5} = 1$, which is often put this way, of course: $\dfrac{x^2}{2^2} - \dfrac{y^2}{(\sqrt{5})^2} = 1$.

And next, a hyperbola has two lines called the <u>asymptotes</u>, the slope of one is **b/a**, and the other is **$-b/a$**. And the asymptotes pass through the center, which is $(0, 0)$ in this case. So the asymptotes are: $y = \pm(b/a)x = \pm\frac{\sqrt{5}}{2}x$.

Next, the <u>eccentricity</u> of a hyperbola is a ratio, <u>the focal distance over half the transverse</u>. So assuming e is the eccentricity, we get: $e = c/a = 3/2$.

Next, a hyperbola has two lines called the <u>directrices</u>, and the distance from each to the center is a ratio, which is <u>half the transverse over the eccentricity</u>.

So since the center is $(0, 0)$, and the hyperbola is horizontal, the directrices are: $x = \pm a/e = \pm a^2/c = \pm 4/3$.

Fig. 0

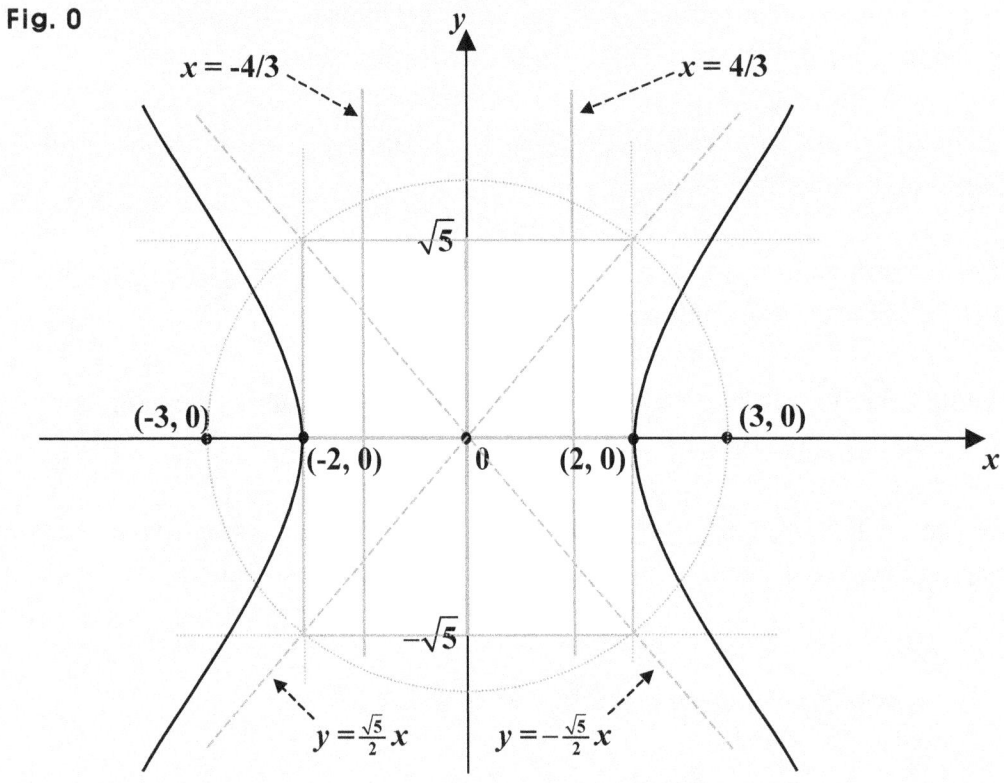

The hyperbola is: $\dfrac{x^2}{4} - \dfrac{y^2}{5} = 1$, often put this way, too, of course: $\dfrac{x^2}{2^2} - \dfrac{y^2}{(\sqrt{5})^2} = 1$.

Note that half the transverse axis is called the semi major axis, too, and half the conjugate axis is called the semi minor axis.

Suggestions or Solutions
To the Problem in the Example 1

Assuming $C(0, 0)$ is the center of a hyperbola, $F(0, 3)$ is a focus, and $V(0, 2)$ is a vertex, find the hyperbola, and its elements, and put them all in a graph.

To begin with, the center is (0, 0), one focus is (0, 3), and the x-coordinates are the same. So the hyperbola is vertical, the other focus is (0, -3), the other vertex is (0, -2), and assuming c is the focal distance, we get: $c = 3$. Then, assuming a is half the conjugate axis, and b is half the transverse axis, we get:

$b = 2$, and $c^2 = a^2 + b^2 \Rightarrow 3^2 = a^2 + 2^2 \Rightarrow a^2 = 5$.

So the transverse axis is **4**, the conjugate axis is $2\sqrt{5}$, and the hyperbola is: $\dfrac{y^2}{4} - \dfrac{x^2}{5} = 1$.

And the asymptotes are: $y = \pm(b/a)x = \pm\frac{2\sqrt{5}}{5}x$.

Next, assuming e is the eccentricity, we get: $e = c/b = 3/2$.

And next, the directrices are: $y = \pm b/e = \pm b^2/c = \pm 4/3$.

If not quite sure of the idea behind the processes above, follow the steps below:

To begin with, the hyperbola we want to find is centered at the origin. And if it is horizontal, the equation is: $\dfrac{x^2}{a^2} - \dfrac{y^2}{b^2} = 1$. If it is vertical, the equation is: $\dfrac{y^2}{b^2} - \dfrac{x^2}{a^2} = 1$.

So if finding if the hyperbola is horizontal or vertical, and the values of a and b, we find the hyperbola.   How then, can we get them?

To begin with, if the hyperbola is <u>horizontal</u>, <u>the coefficient of x^2 is positive</u>.
Then, <u>a is half the transverse axis</u>, and b is half the conjugate axis.

If it is <u>vertical</u> however, <u>the coefficient of y^2 is positive.</u>
Then, a is half the conjugate axis, and <u>b is half the transverse axis</u>.

Next, the center is (0, 0), and one of the foci is (0, 3).
So we can notice that <u>the center and the foci share the same x-coordinate</u>, which is 0.
We can see thus, the hyperbola is <u>vertical</u>.

So first, assuming the other focus is $(0, p)$, since the center is $(0, 0)$, and is the midpoint between the foci, we get: $0 = (p + 3)/2$.   The other focus is thus, $(0, -3)$.

Next, assuming c is the focal distance, we get: $c = 3$, because the focal distance is the distance from the center to a focus.

Next, if the hyperbola is <u>vertical</u>, the center and vertices share the <u>same x-coordinate</u>, which is 0 in this case. And one vertex is $(0, 2)$.

So assuming the other vertex is $(0, q)$, since the center is $(0, 0)$, and is the midpoint between the vertices, too, we get: $0 = (q + 2)/2$.   The other vertex is thus, $(0, -2)$.

Next, we can say that b <u>is half the transverse axis</u>, which is the distance from a vertex to the center, and thus, is 2. So the transverse axis is 4.   What then, about a?

We have: $c^2 = a^2 + b^2$.   So we get: $3^2 = a^2 + 2^2 \Rightarrow a^2 = 9 - 4 = 5$.

And $2a$ is the conjugate axis, which is thus, $2\sqrt{5}$.

So the hyperbola is: $\dfrac{y^2}{4} - \dfrac{x^2}{5} = 1$, which is often put this way, of course: $\dfrac{y^2}{2^2} - \dfrac{x^2}{(\sqrt{5})^2} = 1$.

Next, a hyperbola has two lines called the <u>asymptotes</u>, the slope of one is b/a, and the other is $-b/a$. And the asymptotes pass through the center, which is $(0, 0)$ in this case. So the asymptotes are: $y = \pm(b/a)x = \pm\frac{2}{\sqrt{5}}x = \pm\frac{2\sqrt{5}}{5}x$.

Next, the <u>eccentricity</u> of a hyperbola is a ratio: <u>the focal distance over half the transverse</u>. So assuming e is the eccentricity, we get: $e = c/b = 3/2$.

And next, a hyperbola has two lines called the <u>directrices</u>, and the distance from each to the center is the value of a ratio: <u>half the transverse over the eccentricity</u>.

So since the center is $(0, 0)$, and the hyperbola is vertical, the directrices are: $y = \pm b/e = \pm b^2/c = \pm 4/3$.

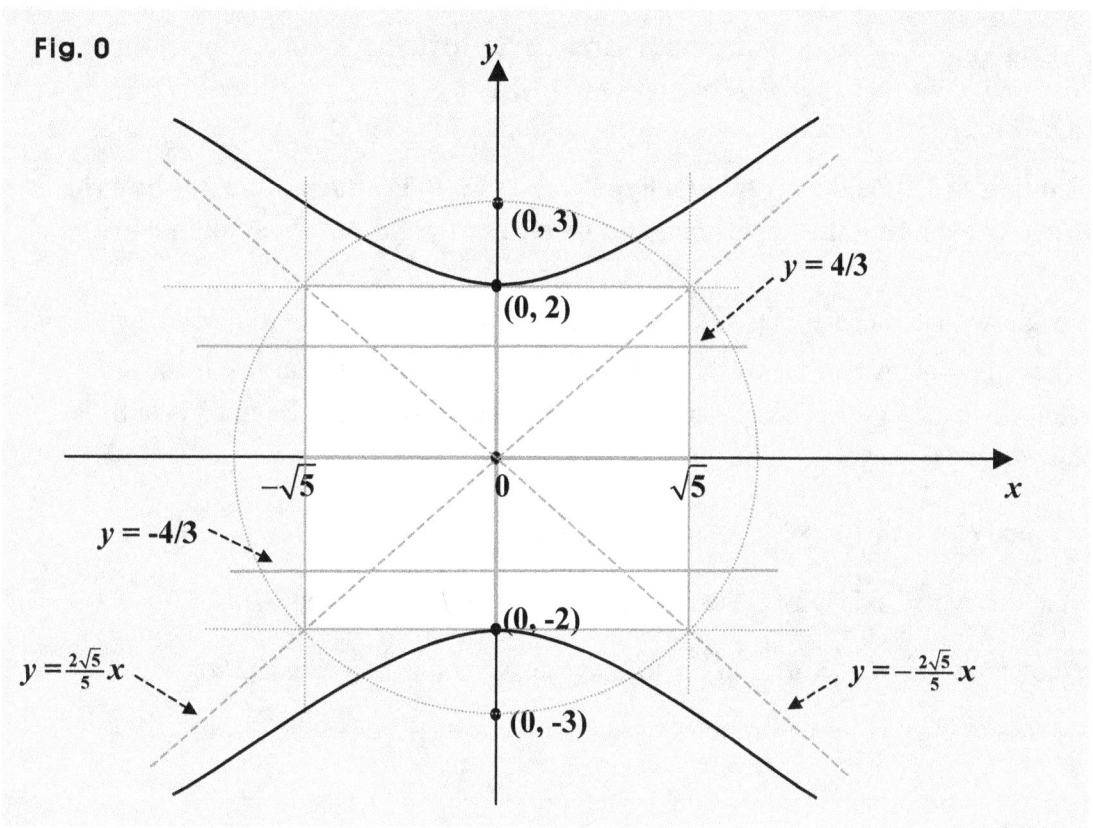

Fig. 0

The hyperbola is: $\dfrac{y^2}{4} - \dfrac{x^2}{5} = 1$, often put this way, too, of course: $\dfrac{y^2}{2^2} - \dfrac{x^2}{(\sqrt{5})^2} = 1$.

Note that half the transverse axis is called the semi major axis, too, and half the conjugate axis is called the semi minor axis.

Suggestions or Solutions
To the Problem in the Example 2

Assuming $C(0, 0)$ is the center of a hyperbola, $F(3, 0)$ is a focus, and 2 is half the conjugate axis, find the hyperbola, and its elements, and put them all in a graph.

To begin with, the center is $(0, 0)$, one focus is $(3, 0)$, and the y-coordinates are the same. So the hyperbola is horizontal, the other focus is $(-3, 0)$, and assuming c is the focal distance, we get: $c = 3$. Then, assuming a is half the transverse axis, and b is half the conjugate axis, we get:

$b = 2$, and $c^2 = a^2 + b^2 \Rightarrow 3^2 = a^2 + 2^2 \Rightarrow a^2 = 5$.

So the transverse axis is $2\sqrt{5}$, the conjugate axis is **4**, and the hyperbola is: $\dfrac{x^2}{5} - \dfrac{y^2}{4} = 1$.

So the vertices are $(-\sqrt{5}, 0)$ and $(\sqrt{5}, 0)$, and the asymptotes are: $y = \pm(b/a)x = \pm\frac{2\sqrt{5}}{5}x$.

Next, assuming e is the eccentricity, we get: $e = c/a = \frac{3\sqrt{5}}{5}$.

And next, the directrices are: $x = \pm a/e = \pm a^2/c = \pm 5/3$.

If not quite sure of the idea behind the processes above, follow the steps below:

To begin with, the hyperbola we want to find is centered at the origin. And if it is horizontal, the equation is: $\dfrac{x^2}{a^2} - \dfrac{y^2}{b^2} = 1$. If it is vertical, the equation is: $\dfrac{y^2}{b^2} - \dfrac{x^2}{a^2} = 1$.

So if finding if the hyperbola is horizontal or vertical, and the values of a and b, we find the hyperbola. How then, can we get them?

To begin with, if the hyperbola is <u>horizontal, the coefficient of x^2 is positive</u>.
Then, <u>**a** is half the transverse axis</u>, and b is half the conjugate axis.

If it is <u>vertical</u> however, <u>the coefficient of y^2 is positive.</u>
Then, a is half the conjugate axis, and <u>**b** is half the transverse axis</u>.

Next, the center is $(0, 0)$, and one of the foci is $(3, 0)$.
So we can notice that <u>the center and the foci share the same y-coordinate</u>, which is 0.
We can see thus, the hyperbola is <u>horizontal</u>.

So first, assuming the other focus is $(p, 0)$, since the center is $(0, 0)$, and is the midpoint between the foci, we get: $0 = (p + 3)/2$.   The other focus is thus, $(-3, 0)$.

Next, assuming c is the focal distance, we get: $c = 3$, because the focal distance is the distance from the center to a focus.

Next, we can say that b is half the conjugate axis, and thus, is 2.
So the conjugate axis is 4.   What then, about a?

We have: $c^2 = a^2 + b^2$.   So we get: $3^2 = a^2 + 2^2 \Rightarrow a^2 = 9 - 4 = 5$.

And a is half the transverse axis, so the transverse axis is $2\sqrt{5}$.

So the hyperbola is: $\dfrac{x^2}{5} - \dfrac{y^2}{4} = 1$, which is often put this way, of course: $\dfrac{x^2}{(\sqrt{5})^2} - \dfrac{y^2}{2^2} = 1$.

Next, if the hyperbola is <u>horizontal</u>, the center and vertices share the <u>same y-coordinate,</u> which is 0 in this case.
And the center is the midpoint between the <u>vertices</u>, too, which are the <u>endpoints</u> of the <u>transverse axis</u>, which is $2a$, and is $2\sqrt{5}$.
And a is half the transverse axis, which is the distance from a vertex to the center.
So since the center is $(0, 0)$, the two vertices are $(-\sqrt{5}, 0)$ and $(\sqrt{5}, 0)$.

And next, a hyperbola has two lines called the <u>asymptotes</u>, the slope of one is b/a, and the other is $-b/a$. And the asymptotes pass through the center, which is $(0, 0)$ in this case.
So the asymptotes are: $y = \pm(b/a)x = \pm\frac{2}{\sqrt{5}}x = \pm\frac{2\sqrt{5}}{5}x$.

Next, the <u>eccentricity</u> of a hyperbola is a ratio, <u>the focal distance over half the transverse.</u>
So assuming e is the eccentricity, we get: $e = c/a = \frac{3}{\sqrt{5}} = \frac{3\sqrt{5}}{5}$.

Next, a hyperbola has two lines called the <u>directrices</u>, and the distance from each to the center is a ratio, which is <u>half the transverse over the eccentricity.</u>

So since the center is $(0, 0)$, and the hyperbola is horizontal, the directrices are:
$x = \pm a/e = \pm a^2/c = \pm 5/3$.

Fig. 0

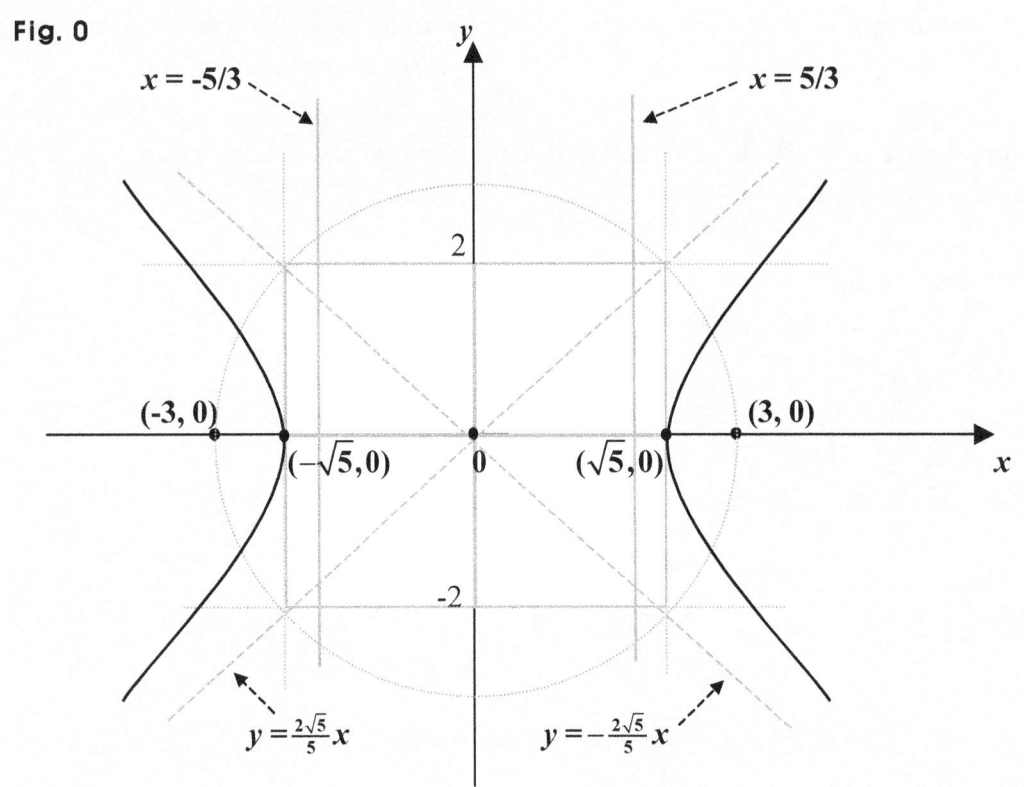

The hyperbola is: $\dfrac{x^2}{5} - \dfrac{y^2}{4} = 1$, which is often put this way, of course: $\dfrac{x^2}{(\sqrt{5})^2} - \dfrac{y^2}{2^2} = 1$.

Note that half the transverse axis is called the semi major axis, too, and half the conjugate axis is called the semi minor axis.

Examples 2 in Hyperbolas

Assuming in each example below, C is the center of a hyperbola, F is a focus, V is a vertex, T is half the transverse axis, and t is half the conjugate axis, find the hyperbola, the eccentricity, the asymptotes, and the directrices, and put in a graph the hyperbola, together with the foci, vertices, asymptotes, and directrices.

0. $C(0, 0)$, $F(0, 3)$, and $t = 2$.

1. $C(0, 1)$, $F(-2, 1)$, and $T = 1$.

2. $C(1, 1)$, $F(1, 6)$, and $t = 4$.

Suggestions or Solutions
To the Problem in the Example 0

Assuming $C(0, 0)$ is the center of a hyperbola, $F(0, 3)$ is a focus, and 2 is half the conjugate axis, find the hyperbola, and its elements, and put them all in a graph.

To begin with, the center is $(0, 0)$, one focus is $(0, 3)$, and the x-coordinates are the same. So the hyperbola is vertical, the other focus is $(0, -3)$, and assuming c is the focal distance, we get: $c = 3$. Then, assuming a is half the conjugate axis, and b is half the transverse axis, we get:

$a = 2$, and $c^2 = a^2 + b^2 \Rightarrow 3^2 = 2^2 + b^2 \Rightarrow b^2 = 5$.

So the transverse axis is $2\sqrt{5}$, the conjugate axis is 4, and the hyperbola is: $\dfrac{y^2}{5} - \dfrac{x^2}{4} = 1$.

So the vertices are $(0, \sqrt{5})$ and $(0, -\sqrt{5})$, and the asymptotes are: $y = \pm(b/a)x = \pm\frac{\sqrt{5}}{2}x$.

Next, assuming e is the eccentricity, we get: $e = c/b = \frac{3\sqrt{5}}{5}$.

And next, the directrices are: $y = \pm b/e = \pm b^2/c = \pm 5/3$.

If not quite sure of the idea behind the processes above, follow the steps below:

To begin with, the hyperbola we want to find is centered at the origin. And if it is horizontal, the equation is: $\dfrac{x^2}{a^2} - \dfrac{y^2}{b^2} = 1$. If it is vertical, the equation is: $\dfrac{y^2}{b^2} - \dfrac{x^2}{a^2} = 1$.

So if finding if the hyperbola is horizontal or vertical, and the values of a and b, we find the hyperbola. How then, can we get them?

To begin with, if the hyperbola is horizontal, the coefficient of x^2 is positive. Then, a is half the transverse axis, and b is half the conjugate axis.

If it is vertical however, the coefficient of y^2 is positive. Then, a is half the conjugate axis, and b is half the transverse axis.

Next, the center is $(0, 0)$, and one of the foci is $(0, 3)$. So we can notice that the center and the foci share the same x-coordinate, which is 0. We can see thus, the hyperbola is vertical.

So first, assuming the other focus is $(0, q)$, since the center is $(0, 0)$, and is the midpoint between the foci, we get: $0 = (q + 3)/2$.   The other focus is thus, $(0, -3)$.

Next, assuming c is the focal distance, we get: $c = 3$, because the focal distance is the distance from the center to a focus.

Next, we can say that a is half the conjugate axis, and thus, is 2.
So the conjugate axis is 4.   What then, about b?

We have: $c^2 = a^2 + b^2$.   So we get: $3^2 = 2^2 + b^2 \Rightarrow b^2 = 9 - 4 = 5$.

And b is half the transverse axis, so the transverse axis is $2\sqrt{5}$.

So the hyperbola is: $\dfrac{y^2}{5} - \dfrac{x^2}{4} = 1$, which is often put this way, of course: $\dfrac{y^2}{(\sqrt{5})^2} - \dfrac{x^2}{2^2} = 1$.

Next, if the hyperbola is <u>vertical</u>, the center and vertices share the <u>same x-coordinate</u>, which is 0 in this case.
And the center is the midpoint between the <u>vertices</u>, too, which are the <u>endpoints</u> of the <u>transverse axis</u>, which is $2b$, and is $2\sqrt{5}$.
And b is half the transverse axis, which is the distance from a vertex to the center.
So since the center is $(0, 0)$, the two vertices are $(0, \sqrt{5})$ and $(0, -\sqrt{5})$.

And next, a hyperbola has two lines called the <u>asymptotes</u>, the slope of one is b/a, and the other is $-b/a$. And the asymptotes pass through the center, which is $(0, 0)$ in this case.
So the asymptotes are: $y = \pm(b/a)x = \pm\frac{\sqrt{5}}{2}x$.

Next, the <u>eccentricity</u> of a hyperbola is a ratio, <u>the focal distance over half the transverse</u>.
So assuming e is the eccentricity, we get: $e = c/b = \frac{3}{\sqrt{5}} = \frac{3\sqrt{5}}{5}$.

Next, a hyperbola has two lines called the <u>directrices</u>, and the distance from each to the center is a ratio, which is <u>half the transverse over the eccentricity</u>.

So since the center is $(0, 0)$, and the hyperbola is vertical, the directrices are:
$y = \pm b/e = \pm b^2/c = \pm 5/3$.

Fig. 0

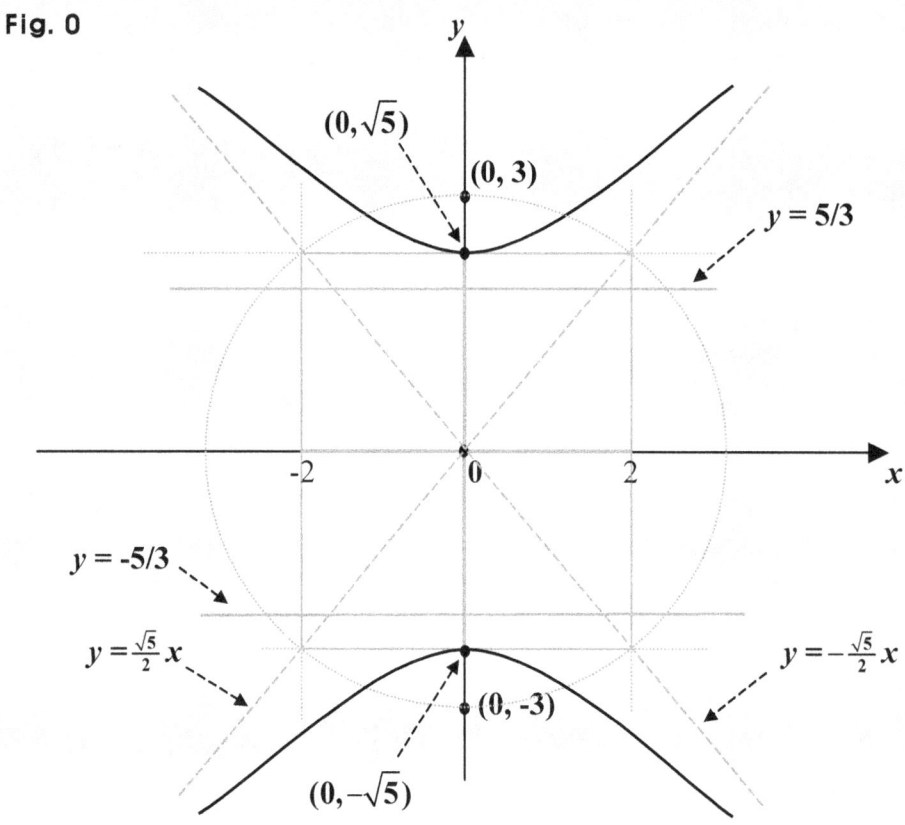

The hyperbola is: $\dfrac{y^2}{5} - \dfrac{x^2}{4} = 1$, often put this way, too, of course: $\dfrac{y^2}{(\sqrt{5})^2} - \dfrac{x^2}{2^2} = 1$.

Note that half the transverse axis is called the semi major axis, too, and half the conjugate axis is called the semi minor axis.

Suggestions or Solutions
To the Problem in the Example 1

Assuming $C(0, 1)$ is the center of a hyperbola, $F(-2, 1)$ is a focus, and 1 is half the transverse axis, find the hyperbola, and its elements, and put them all in a graph.

To begin with, the center is (0, 1), a focus is (-2, 1), and the y-coordinates are the same. So the hyperbola is horizontal, the other focus is (2, 1), and assuming c is the focal distance, we get: $c = 2$. Then, assuming a is half the transverse axis, and b is half the conjugate axis, we get:

$a = 1$, and $c^2 = a^2 + b^2 \Rightarrow 2^2 = 1^2 + b^2 \Rightarrow b^2 = 3$.   So the transverse axis is **2**, the

conjugate axis is $2\sqrt{3}$, and the hyperbola is: $x^2 - \dfrac{(y-1)^2}{3} = 1$.

So the vertices are (-1, 1) and (1, 1), and the asymptotes are: $y = \pm\sqrt{3}x + 1$.

Next, assuming e is the eccentricity, we get: $e = c/a = 2$.

And next, the directrices are: $x = \pm a/e = \pm a^2/c = \pm 1/2$.

If not quite sure of the idea behind the processes above, follow the steps below:

To begin with, the hyperbola we want to find is centered at (0, 1). And if it is horizontal,

the equation is: $\dfrac{x^2}{a^2} - \dfrac{(y-1)^2}{b^2} = 1$. If it is vertical, the equation is: $\dfrac{(y-1)^2}{b^2} - \dfrac{x^2}{a^2} = 1$.

So if finding if the hyperbola is horizontal or vertical, and the values of a and b, we find the hyperbola.   How then, can we get them?

To begin with, if the hyperbola is <u>horizontal, the coefficient of x^2 is positive</u>.
Then, <u>**a** is half the transverse axis</u>, and b is half the conjugate axis.

If it is <u>vertical</u> however, <u>the coefficient of y^2 is positive.</u>
Then, a is half the conjugate axis, and <u>**b** is half the transverse axis</u>.

Next, the center is (0, 1), and one of the foci is (-2, 1).

So we can notice that <u>the center and the foci share the same y-coordinate</u>, which is 1.

We can see thus, the hyperbola is <u>horizontal</u>.

So first, assuming the other focus is (*p*, 1), since the center is (0, 1), and is the midpoint between the foci, we get: $0 = \{p + (-2)\}/2$.   The other focus is thus, (2, 1).

Next, assuming *c* is the focal distance, we get: $c = 2$, because the focal distance is the distance from the center to a focus.

Next, we can say that *a* is half the transverse axis, and thus, is 1.
So the transverse axis is 2.   What then, about *b*?

We have: $c^2 = a^2 + b^2$.   So we get: $2^2 = 1^2 + b^2 \Rightarrow b^2 = 4 - 1 = 3$.

And *b* is half the conjugate axis, so the conjugate axis is $2\sqrt{3}$.

So the hyperbola is: $x^2 - \dfrac{y^2}{3} = 1$, which is often put this way, of course: $\dfrac{x^2}{1^2} - \dfrac{y^2}{(\sqrt{3})^2} = 1$.

Next, if the hyperbola is <u>horizontal</u>, the center and vertices share the <u>same *y*-coordinate</u>, which is 1 in this case.
And the center is the midpoint between the <u>vertices</u>, too, which are the <u>endpoints</u> of the <u>transverse axis</u>, which is **2a**, and is 2.
And *a* is half the transverse axis, which is the distance from a vertex to the center.
So since the center is (0, 1), the two vertices are (-1, 1) and (1, 1).

And next, a hyperbola has two lines called the <u>asymptotes</u>, the slope of one is **b/a**, and the other is **–b/a**. And the asymptotes pass through the center, which is (0, 1) in this case.
And a line of slope **m** passing through (**s, t**) is: $y - t = m(x - s)$.
So getting the asymptotes, we get: $y - 1 = \pm(b/a)x = \pm\frac{\sqrt{3}}{1}x = \pm\sqrt{3}x \Rightarrow y = \pm\sqrt{3}x + 1$.

Next, the <u>eccentricity</u> of a hyperbola is a ratio, <u>the focal distance over half the transverse</u>.
So assuming *e* is the eccentricity, we get: $e = c/a = 2/1 = 2$.

Next, a hyperbola has two lines called the <u>directrices</u>, and the distance from each to the center is the value of a ratio, which is <u>half the transverse over the eccentricity</u>.

So since the center is (0, 1), and the hyperbola is horizontal, the directrices are:
$x = \pm a/e = \pm a^2/c = \pm 1/2$.

Fig. 0

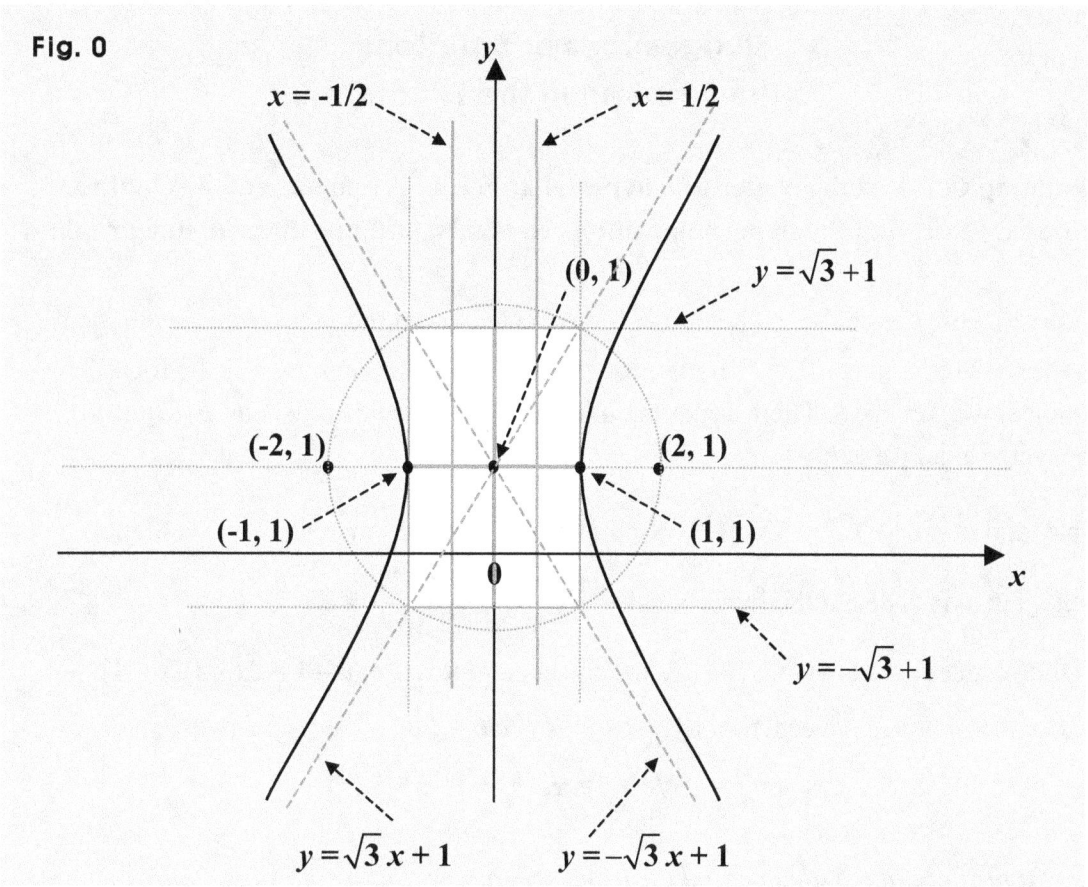

The hyperbola is: $x^2 - \dfrac{(y-1)^2}{3} = 1$, often put this way, too, of course: $\dfrac{x^2}{1^2} - \dfrac{(y-1)^2}{(\sqrt{3})^2} = 1$.

Note that half the transverse axis is called the semi major axis, too, and half the conjugate axis is called the semi minor axis.

Suggestions or Solutions
To the Problem in the Example 2

Assuming $C(1, 1)$ is the center of a hyperbola, $F(1, 6)$ is a focus, and 4 is half the conjugate axis, find the hyperbola, and its elements, and put them all in a graph.

To begin with, the center is (1, 1), one focus is (1, 6), and the x-coordinates are the same. So the hyperbola is vertical, the other focus is (1, -4), and assuming c is the focal distance, we get: $c = 5$. Then, assuming a is half the conjugate axis, and b is half the transverse axis, we get:

$a = 4$, and $c^2 = a^2 + b^2 \Rightarrow 5^2 = 4^2 + b^2 \Rightarrow b^2 = 3^2$.   So the transverse axis is **6**, the conjugate axis is **8**, and the hyperbola is: $\dfrac{(y-1)^2}{9} - \dfrac{(x-1)^2}{16} = 1$.

So the vertices are (1, 4) and (1, -2), and the asymptotes are: $y - 1 = \pm(3/4)(x - 1)$.

Next, assuming e is the eccentricity, we get: $e = c/b = 5/4$.

And next, the directrices are: $y = \pm b/e + 1 = \pm b^2/c + 1 = \pm 9/5 + 1$.

If not quite sure of the idea behind the processes above, follow the steps below:

To begin with, the hyperbola we want to find is centered at (1, 1). And if it is horizontal, the equation is: $\dfrac{(x-1)^2}{a^2} - \dfrac{(y-1)^2}{b^2} = 1$. If vertical, the equation is: $\dfrac{(y-1)^2}{b^2} - \dfrac{(x-1)^2}{a^2} = 1$.

So if finding if the hyperbola is horizontal or vertical, and the values of a and b, we find the hyperbola.   How then, can we get them?

To begin with, if the hyperbola is <u>horizontal, the coefficient of x^2 is positive.</u>
Then, a <u>is half the transverse axis</u>, and b is half the conjugate axis.

If it is <u>vertical</u> however, <u>the coefficient of y^2 is positive.</u>
Then, a is half the conjugate axis, and b <u>is half the transverse axis.</u>

Next, the center is (1, 1), and one of the foci is (1, 6).
So we can notice that <u>the center and the foci share the same x-coordinate</u>, which is 1.
We can see thus, the hyperbola is <u>vertical</u>.

So first, assuming the other focus is $(1, q)$, since the center is $(1, 1)$, and is the midpoint between the foci, we get: $1 = (q + 6)/2$.   The other focus is thus, $(1, -4)$.

Next, assuming c is the focal distance, we get: $c = 5$, because the focal distance is the distance from the center to a focus.

Next, we can say that a is half the conjugate axis, and thus, is 4.
So the conjugate axis is 8.   What then, about b?

We have: $c^2 = a^2 + b^2$.   So we get: $5^2 = 4^2 + b^2 \Rightarrow b^2 = 25 - 16 = 9$.

And b is half the transverse axis, so the transverse axis is 6.

So the hyperbola is: $\dfrac{(y-1)^2}{9} - \dfrac{(x-1)^2}{16} = 1$, often put this way: $\dfrac{(y-1)^2}{3^2} - \dfrac{(x-1)^2}{4^2} = 1$.

Next, if the hyperbola is <u>vertical</u>, the center and vertices share the <u>same x-coordinate</u>, which is 1 in this case.
And the center is the midpoint between the <u>vertices</u>, too, which are the <u>endpoints</u> of the <u>transverse axis</u>, which is $2b$, and is 6.
And b is half the transverse axis, which is the distance from a vertex to the center.
So since the center is $(1, 1)$, the two vertices are $(1, 4)$ and $(1, -2)$

And next, a hyperbola has two lines called the <u>asymptotes</u>, the slope of one is b/a, and the other is $-b/a$. And the asymptotes pass through the center, which is $(1, 1)$ in this case.
And a line of slope m passing through (s, t) is: $y - t = m(x - s)$.
So getting the asymptotes, we get: $y - 1 = \pm(b/a)(x - 1) \Rightarrow y - 1 = \pm(3/4)(x - 1)$.

Next, the <u>eccentricity</u> of a hyperbola is a ratio, <u>the focal distance over half the transverse</u>.
So assuming e is the eccentricity, we get: $e = c/b = 5/3$.

Next, a hyperbola has two lines called the <u>directrices</u>, and the distance from each to the center is a ratio, which is <u>half the transverse over the eccentricity</u>.

So since the center is $(1, 1)$, and the hyperbola is vertical, the directrices are:
$y = \pm b/e + 1 = \pm b^2/c + 1 = \pm 5/3 + 1$.

Fig. 0

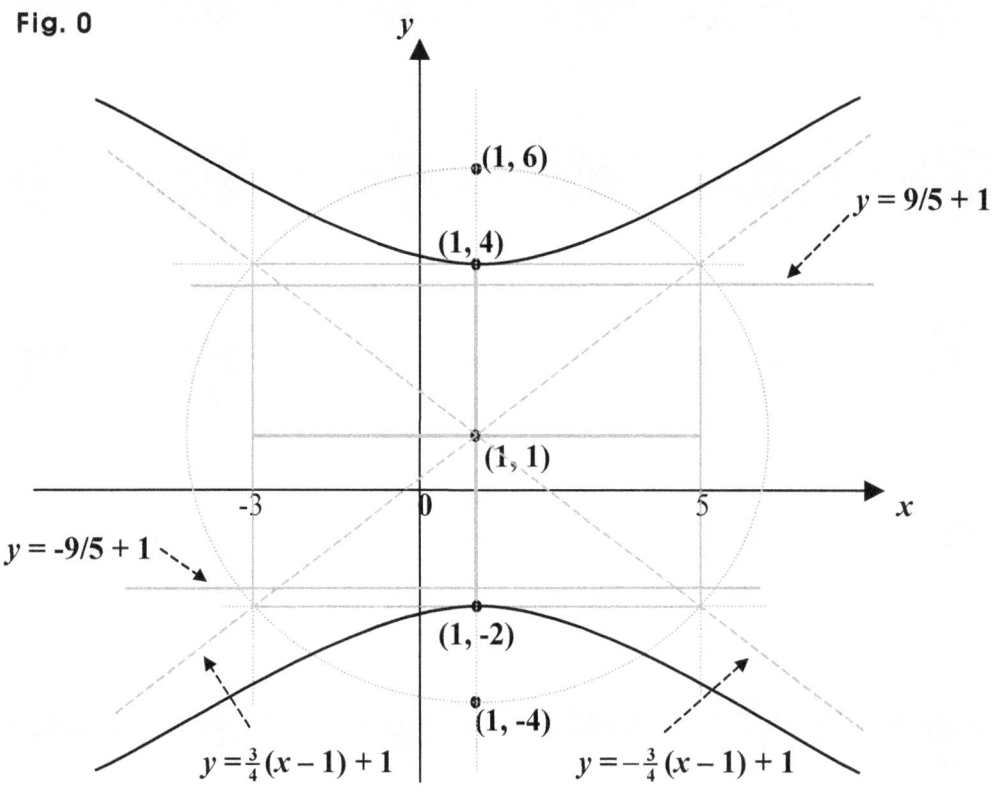

The hyperbola is: $\dfrac{y^2}{5} - \dfrac{x^2}{4} = 1,$ often put this way, too, of course: $\dfrac{y^2}{(\sqrt{5})^2} - \dfrac{x^2}{2^2} = 1.$

Note that half the transverse axis is called the semi major axis, too, and half the conjugate axis is called the semi minor axis.

Examples 3 in Hyperbolas

Find if each hyperbola below is horizontal or vertical, the center, foci, vertices, transverse axis, conjugate axis, asymptotes, eccentricity, and directrices.

0. $\dfrac{x^2}{25} - \dfrac{y^2}{16} = 1$

1. $\dfrac{y^2}{25} - \dfrac{x^2}{16} = 1$

2. $\dfrac{(x-2)^2}{16} - \dfrac{(y-1)^2}{25} = 1$

Suggestions or Solutions
To the Problem in the Example 0

Find all the elements of the hyperbola as follows: $\dfrac{x^2}{25} - \dfrac{y^2}{16} = 1$.

To begin with, the hyperbola is horizontal, the center is (0, 0), the transverse axis is 10, and the conjugate axis is 8.

Next, assuming c is the focal distance, a is half the transverse axis, and b is half the conjugate axis, we get: $a = 5$, $b = 4$, and $c^2 = a^2 + b^2 \Rightarrow c^2 = 25 + 16 = 41 \Rightarrow c = \sqrt{41}$.

So the focal distance is $\sqrt{41}$, and the foci are $(-\sqrt{41}, 0)$ and $(\sqrt{41}, 0)$.
And the vertices are (-5, 0) and (5, 0).

Next, the asymptotes are: $y = \pm(b/a)x = \pm\frac{4}{5}x$.

Next, assuming e is the eccentricity, we get: $e = c/a = \frac{\sqrt{41}}{5}$.

And next, the directrices are: $x = \pm a/e = \pm a^2/c = \pm\frac{25\sqrt{41}}{41}$.

If not quite sure of the idea behind the solution above, follow the steps below:

To begin with, if a hyperbola is centered at the origin, and is horizontal, the equation is:
$\dfrac{x^2}{a^2} - \dfrac{y^2}{b^2} = 1$, where a is half the transverse axis, and b is half the conjugate axis.

If vertical though, its equation is: $\dfrac{y^2}{b^2} - \dfrac{x^2}{a^2} = 1$, where b is half the transverse axis, and a is half the conjugate axis.

Next, we can put the hyperbola given this way: $\dfrac{x^2}{5^2} - \dfrac{y^2}{4^2} = 1$.

So the hyperbola given is horizontal, the center is (0, 0), the transverse axis is 10, and the conjugate axis is 8.   What then, about the focal distance?

If c is the focal distance, $2a$ is the transverse axis, and $2b$ is the conjugate axis, we get: $c^2 = a^2 + b^2$.   So we get: $c^2 = 25 + 16 = 41 \Rightarrow c = \sqrt{41}$.

Next, the center, foci, and vertices are all in the transverse axis. So if the hyperbola is horizontal, the center, foci, and vertices share the same y-coordinate.

The center is the midpoint between the foci, and the focal distance is the distance from the center to each focus, and is c, which is $\sqrt{41}$.

And also, the center is the midpoint between the vertices, too, which are the endpoints of the transverse axis, which is twice the distance from the center to each vertex, and the distance in this case, is a, which is 5.

So since the center is (0, 0), and the hyperbola is horizontal, the foci are $(-\sqrt{41},0)$ and $(\sqrt{41},0)$, and the vertices are **(-5, 0)** and **(5, 0)**.

Next, a hyperbola has two lines called the asymptotes, the slope of one is b/a, and the other is $-b/a$. And the asymptotes pass through the center, which is (0, 0) in this case. So the asymptotes are: $y = \pm(b/a)x = \pm\frac{4}{5}x$.

Next, the eccentricity of a hyperbola is a ratio, the focal distance over half the transverse axis. So assuming e is the eccentricity, we get: $e = c/a = \frac{\sqrt{41}}{5}$.

And next, a hyperbola has two lines called the directrices, and the distance from each to the center is a ratio, half the transverse over the eccentricity. So since the center is (0, 0), and the hyperbola is horizontal, the directrices are: $x = \pm a/e = \pm a^2/c = \pm\frac{25}{\sqrt{41}} = \pm\frac{25\sqrt{41}}{41}$.

Fig. 0

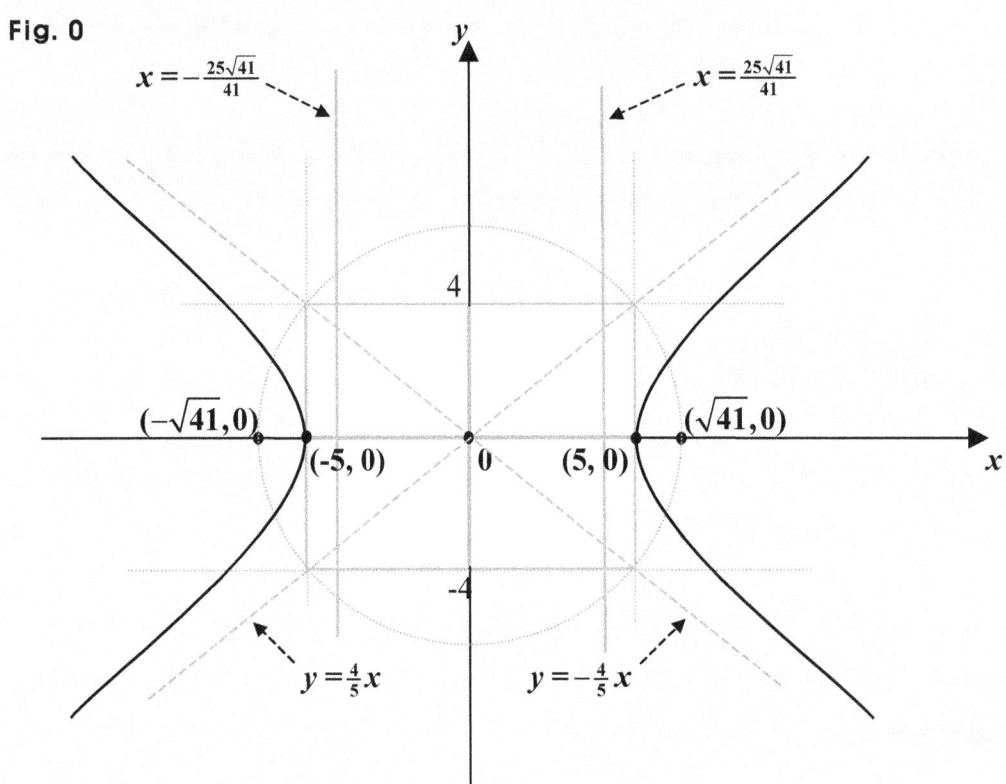

The hyperbola is: $\dfrac{x^2}{25} - \dfrac{y^2}{16} = 1$, often put this way, too, of course: $\dfrac{x^2}{5^2} - \dfrac{y^2}{4^2} = 1$.

Note that half the transverse axis is called the semi major axis, too, and half the conjugate axis is called the semi minor axis.

Suggestions or Solutions
To the Problem in the Example 1

Find all the elements of the hyperbola as follows: $\dfrac{y^2}{25} - \dfrac{x^2}{16} = 1$.

To begin with, the hyperbola is vertical, the center is $(0, 0)$, the transverse axis is 10, and the conjugate axis is 8.

Next, assuming c is the focal distance, b is half the transverse axis, and a is half the conjugate axis, we get: $b = 5$, $a = 4$, and $c^2 = a^2 + b^2 \Rightarrow c^2 = 16 + 25 = 41 \Rightarrow c = \sqrt{41}$.

So the focal distance is $\sqrt{41}$, and the foci are $(0, \sqrt{41})$ and $(0, -\sqrt{41})$.
And the vertices are $(0, 5)$ and $(0, -5)$.

Next, the asymptotes are: $y = \pm(b/a)x = \pm\frac{5}{4}x$.

Next, assuming e is the eccentricity, we get: $e = c/b = \frac{\sqrt{41}}{5}$.

And next, the directrices are: $y = \pm b/e = \pm b^2/c = \pm\frac{25\sqrt{41}}{41}$.

If not quite sure of the idea behind the solution above, follow the steps below:

To begin with, if a hyperbola is centered at the origin, and is horizontal, the equation is: $\dfrac{x^2}{a^2} - \dfrac{y^2}{b^2} = 1$, where a is half the transverse axis, and b is half the conjugate axis.

If it is vertical, the equation is: $\dfrac{y^2}{b^2} - \dfrac{x^2}{a^2} = 1$, where b is half the transverse axis, and a is half the conjugate axis.

Next, we can put the hyperbola given this way: $\dfrac{y^2}{5^2} - \dfrac{x^2}{4^2} = 1$.

So the hyperbola given is vertical, the center is $(0, 0)$, the transverse axis is 10, and the conjugate axis is 8.   What then, about the focal distance?

If c is the focal distance, $2b$ is the transverse axis, and $2a$ is the conjugate axis, we get: $c^2 = a^2 + b^2$.   So we get: $c^2 = 16 + 25 = 41 \Rightarrow c = \sqrt{41}$.

Next, the center, foci, and vertices are all in the transverse axis. So if the hyperbola is vertical, the center, foci, and vertices share the same *x*-coordinate.

The center is the midpoint between the foci, and the focal distance is the distance from the center to each focus, and is *c*, which is $\sqrt{41}$.

And also, the center is the midpoint between the vertices, too, which are the endpoints of the transverse axis, which is twice the distance from the center to each vertex, and the distance in this case, is *b*, which is 5.

So since the center is (0, 0), and the hyperbola is vertical, the foci are $(0, \sqrt{41})$ and $(0, -\sqrt{41})$, and the vertices are **(0, 5)** and **(0, -5)**.

Next, a hyperbola has two lines called the asymptotes, the slope of one is *b/a*, and the other is *−b/a*. And the asymptotes pass through the center, which is (0, 0) in this case. So the asymptotes are: $y = \pm(b/a)x = \pm\frac{5}{4}x$.

Next, the eccentricity of a hyperbola is a ratio, the focal distance over half the transverse axis. So assuming *e* is the eccentricity, we get: $e = c/b = \frac{\sqrt{41}}{5}$.

And next, a hyperbola has two lines called the directrices, and the distance from each to the center is a ratio, half the transverse over the eccentricity. So since the center is (0, 0), and the hyperbola is vertical, the directrices are: $y = \pm b/e = \pm b^2/c = \pm\frac{25}{\sqrt{41}} = \pm\frac{25\sqrt{41}}{41}$.

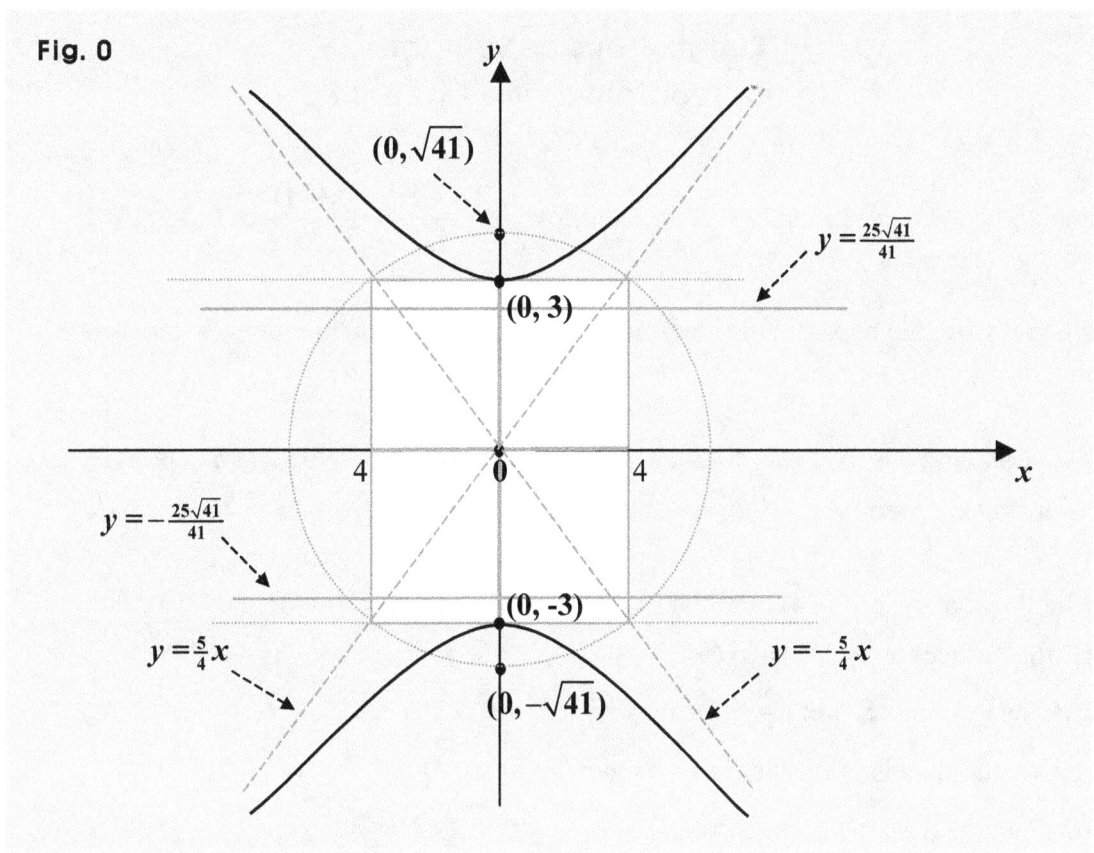

Fig. 0

The hyperbola is: $\dfrac{y^2}{25} - \dfrac{x^2}{16} = 1$, often put this way, too, of course: $\dfrac{y^2}{5^2} - \dfrac{x^2}{4^2} = 1$.

Note that half the transverse axis is called the semi major axis, too, and half the conjugate axis is called the semi minor axis.

Suggestions or Solutions
To the Problem in the Example 2

Find all the elements of the hyperbola as follows: $\dfrac{(x-2)^2}{16} - \dfrac{(y-1)^2}{25} = 1$.

To begin with, the hyperbola is horizontal, the center is (2, 1), the transverse axis is 8, and the conjugate axis is 10.

Next, assuming c is the focal distance, a is half the transverse axis, and b is half the conjugate axis, we get: $a = 4$, $b = 5$, and $c^2 = a^2 + b^2 \Rightarrow c^2 = 16 + 25 = 41 \Rightarrow c = \sqrt{41}$.

So the focal distance is $\sqrt{41}$, and the foci are $(2 - \sqrt{41}, 1)$ and $(2 + \sqrt{41}, 1)$.
And the vertices are (-2, 1) and (6, 1).

Next, the asymptotes are: $y = \pm(b/a)x = \pm\frac{5}{4}x$.

Next, assuming e is the eccentricity, we get: $e = c/a = \frac{\sqrt{41}}{4}$.

And next, the directrices are: $x = \pm a/e + 2 = \pm a^2/c + 2 = \pm \frac{16\sqrt{41}}{41} + 2$.

If not quite sure of the idea behind the solution above, follow the steps below:

To begin with, if a hyperbola is centered at (u, v), and is horizontal, the equation is:
$\dfrac{(x-u)^2}{a^2} - \dfrac{(y-v)^2}{b^2} = 1$, where a is half the transverse axis, and b is half the conjugate

axis. And if vertical, its equation is: $\dfrac{(y-v)^2}{b^2} - \dfrac{(x-u)^2}{a^2} = 1$, where b is half the

transverse axis, and a is half the conjugate axis.

Next, we can put the hyperbola given this way: $\dfrac{(x-2)^2}{4^2} - \dfrac{(y-1)^2}{5^2} = 1$.

So the hyperbola given is horizontal, the center is (2, 1), the transverse axis is 8, and the conjugate axis is 10.　　What then, about the focal distance?

If c is the focal distance, $2a$ is the transverse axis, and $2b$ is the conjugate axis, we get:
$c^2 = a^2 + b^2$.　　So we get: $c^2 = 16 + 25 = 41 \Rightarrow c = \sqrt{41}$.

Next, the center, foci, and vertices are all in the transverse axis. So if the hyperbola is horizontal, the center, foci, and vertices share the <u>same y-coordinate</u>.

The center is the midpoint between the foci, and the focal distance is the distance from the center to each focus, and is c, which is $\sqrt{41}$.

And also, the center is the midpoint between the <u>vertices</u>, too, which are the <u>endpoints</u> of the <u>transverse axis</u>, which is twice the distance from the center to each vertex, and the distance in this case, is a, which is 4.

So since the center is (2, 1), and the hyperbola is horizontal, the foci are $(2-\sqrt{41}, 0+1)$ and $(2+\sqrt{41}, 0+1)$, that is, $(2-\sqrt{41}, 1)$ and $(2+\sqrt{41}, 1)$, and the vertices are (-4 + 2, 0 + 1) and (4 + 2, 0 + 1), that is, **(-2, 1)** and **(6, 1)**.

Next, a hyperbola has two lines called the <u>asymptotes</u>, the slope of one is b/a, and the other is $-b/a$. And the asymptotes pass through the center, which is (2, 1) in this case. So the asymptotes are: $y - 1 = \pm(b/a)(x - 2) \Rightarrow y - 1 = \pm\frac{5}{4}(x - 2)$.

Next, the <u>eccentricity</u> of a hyperbola is a ratio, <u>the focal distance over half the transverse axis</u>. So assuming e is the eccentricity, we get: $e = c/a = \frac{\sqrt{41}}{4}$.

And next, a hyperbola has two lines called the <u>directrices</u>, and the distance from each to the center is a ratio, <u>half the transverse over the eccentricity</u>. So since the center is (2, 1), and the hyperbola is <u>horizontal</u>, the directrices are as follows:

$$x = \pm a/e + 2 = \pm a^2/c + 2 = \pm\frac{16}{\sqrt{41}} + 2 = \pm\frac{16\sqrt{41}}{41} + 2.$$

Fig. 0

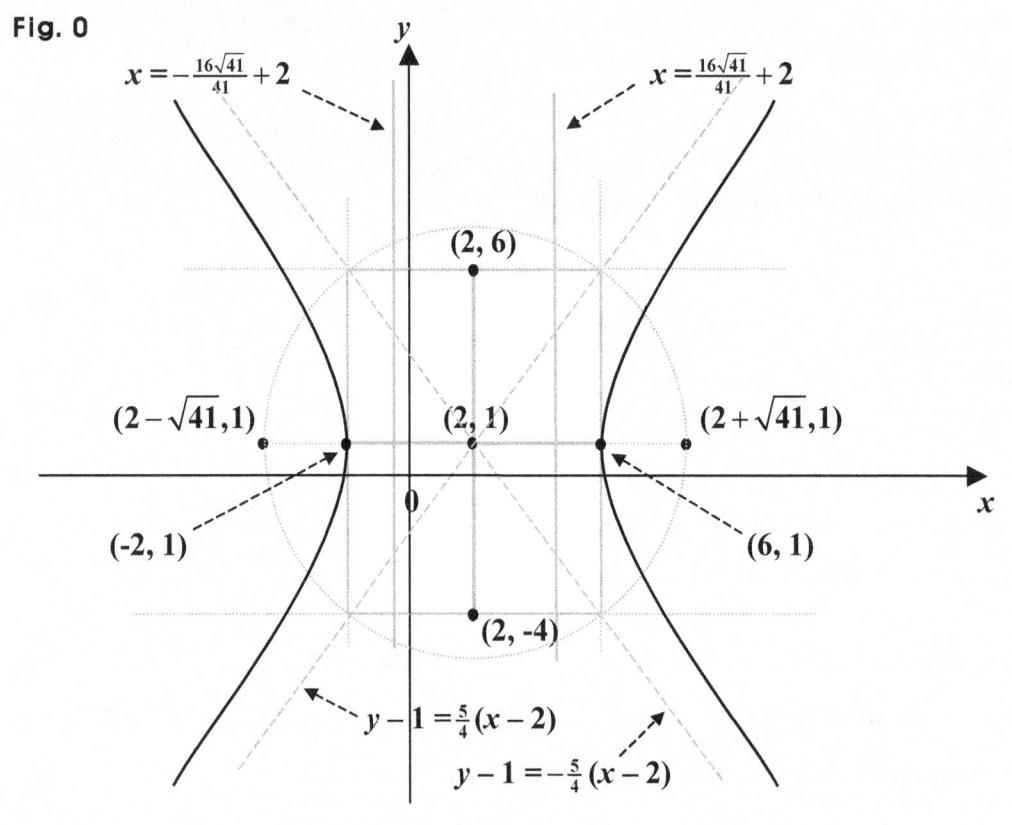

$$x = -\frac{16\sqrt{41}}{41} + 2 \qquad x = \frac{16\sqrt{41}}{41} + 2$$

$(2, 6)$

$(2 - \sqrt{41}, 1)$ $(2, 1)$ $(2 + \sqrt{41}, 1)$

$(-2, 1)$ $(6, 1)$

$(2, -4)$

$$y - 1 = \tfrac{5}{4}(x - 2)$$

$$y - 1 = -\tfrac{5}{4}(x - 2)$$

The hyperbola is: $\dfrac{(x-2)^2}{16} - \dfrac{(y-1)^2}{25} = 1$, often put this way: $\dfrac{(x-2)^2}{4^2} - \dfrac{(y-1)^2}{5^2} = 1$.

Note that half the transverse axis is called the semi major axis, too, and half the conjugate axis is called the semi minor axis.

Examples 4 in Hyperbolas

Find if each hyperbola below is horizontal or vertical, the center, foci, vertices, transverse axis, conjugate axis, asymptotes, eccentricity, and directrices.

0. $25(y-2)^2 - 16(x-1)^2 = 400$

1. $\dfrac{(x-1)^2}{4} - y^2 = 1$

2. $4(x-2)^2 - (y-1)^2 = -4$

3. $1 + 16(y-1)^2 - 25(x-2)^2 = 0$

Suggestions or Solutions
To the Problem in the Example 0

Find all the elements of the hyperbola as follows: $25(y-2)^2 - 16(x-1)^2 = 400$.

To begin with, we can put the hyperbola given this way: $\dfrac{(y-2)^2}{4^2} - \dfrac{(x-1)^2}{5^2} = 1$.

So the hyperbola is vertical, the center is (1, 2), the transverse axis is 8, and the conjugate axis is 10.

Next, assuming c is the focal distance, b is half the transverse axis, and a is half the conjugate axis, we get: $b = 4$, $a = 5$, and $c^2 = a^2 + b^2 \Rightarrow c^2 = 25 + 16 = 41 \Rightarrow c = \sqrt{41}$.

So the focal distance is $\sqrt{41}$, and the foci are $(1, 2+\sqrt{41})$ and $(1, 2-\sqrt{41})$.
And the vertices are (1, 6) and (1, -2).

Next, the asymptotes are: $y - 2 = \pm(b/a)(x-1) \Rightarrow y - 2 = \pm\frac{4}{5}(x-1)$.

Next, assuming e is the eccentricity, we get: $e = c/b = \frac{\sqrt{41}}{4}$.

And next, the directrices are: $y = \pm b/e + 2 = \pm b^2/c + 2 = \pm\frac{16\sqrt{41}}{41} + 2$.

If not quite sure of the idea behind the solution above, follow the steps below:

To begin with, if a hyperbola is centered at (u, v), and is horizontal, the equation is:
$\dfrac{(x-u)^2}{a^2} - \dfrac{(y-v)^2}{b^2} = 1$, where a is half the transverse axis, and b is half the conjugate

axis. And if vertical, its equation is: $\dfrac{(y-v)^2}{b^2} - \dfrac{(x-u)^2}{a^2} = 1$, where b is half the

transverse axis, and a is half the conjugate axis.

Next, we can put the hyperbola given the way below:

$$25(y-2)^2 - 16(x-1)^2 = 400 \Rightarrow \frac{25}{400}(y-2)^2 - \frac{16}{400}(x-1)^2 = 1$$

$$\Rightarrow \frac{5^2}{20^2}(y-2)^2 - \frac{4^2}{20^2}(x-1)^2 \Rightarrow \frac{1}{16}(y-2)^2 - \frac{1}{25}(x-1)^2 = 1 \Rightarrow \frac{(y-2)^2}{4^2} - \frac{(x-1)^2}{5^2} = 1.$$

So the hyperbola given is vertical, the center is (1, 2), the transverse axis is 8, and the conjugate axis is 10.   What then, about the focal distance?

If c is the focal distance, $2b$ is the transverse axis, and $2a$ is the conjugate axis, we get: $c^2 = a^2 + b^2$.   So we get: $c^2 = 25 + 16 = 41 \Rightarrow c = \sqrt{41}$.

Next, the center, foci, and vertices are all in the transverse axis. So if the hyperbola is vertical, the center, foci, and vertices share the same x-coordinate.

The center is the midpoint between the foci, and the focal distance is the distance from the center to each focus, and is c, which is $\sqrt{41}$.

And also, the center is the midpoint between the vertices, too, which are the endpoints of the transverse axis, which is twice the distance from the center to each vertex, and the distance in this case, is b, which is 4.

So since the center is (1, 2), and the hyperbola is vertical, the foci are $(0 + 1, \sqrt{41} + 2)$ and $(0 + 1, 2 - \sqrt{41})$, that is, $(1, \sqrt{41} + 2)$ and $(1, 2 - \sqrt{41})$, and the vertices are $(0 + 1, 4 + 2)$ and $(0 + 1, -4 + 2)$, that is, **(1, 6)** and **(1, -2)**.

Next, a hyperbola has two lines called the asymptotes, the slope of one is b/a, and the other is $-b/a$. And the asymptotes pass through the center, which is (1, 2) in this case. So the asymptotes are: $y - 2 = \pm(b/a)(x - 1) \Rightarrow y - 2 = \pm\frac{5}{4}(x - 1)$.

Next, the eccentricity of a hyperbola is a ratio, the focal distance over half the transverse axis. So assuming e is the eccentricity, we get: $e = c/b = \frac{\sqrt{41}}{4}$.

And next, a hyperbola has two lines called the directrices, and the distance from each to the center is a ratio, half the transverse over the eccentricity.
So since the center is (1, 2), and the hyperbola is vertical, the directrices are as follows:

$$y = \pm b/e + 2 = \pm b^2/c + 2 = \pm\frac{16}{\sqrt{41}} + 2 = \pm\frac{16\sqrt{41}}{41} + 2.$$

Fig. 0

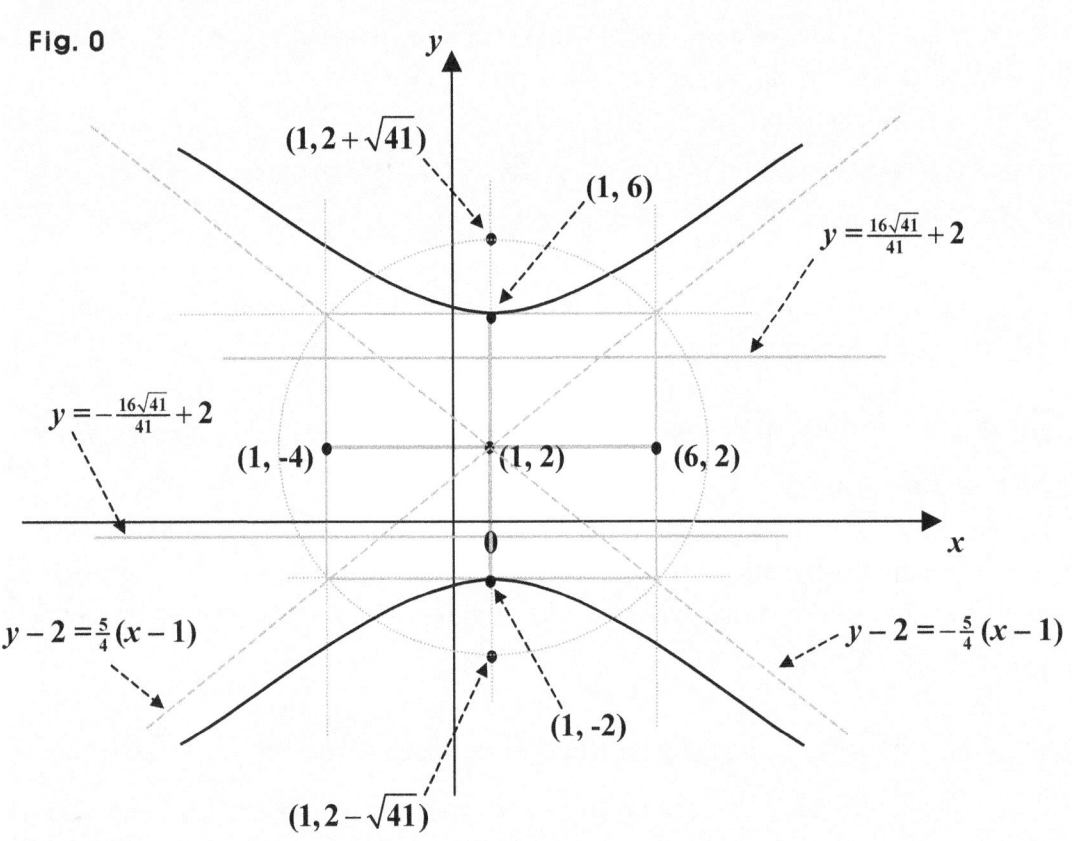

The hyperbola is: $\dfrac{(y-2)^2}{16} - \dfrac{(x-1)^2}{25} = 1$, often put this way: $\dfrac{(y-2)^2}{4^2} - \dfrac{(x-1)^2}{5^2} = 1$.

Note that half the transverse axis is called the semi major axis, too, and half the conjugate axis is called the semi minor axis.

Suggestions or Solutions
To the Problem in the Example 1

Find all the elements of the hyperbola as follows: $\dfrac{(x-1)^2}{4} - y^2 = 1$.

To begin with, the hyperbola is horizontal, the center is (1, 0), the transverse axis is 4, and the conjugate axis is 2.

Next, assuming c is the focal distance, a is half the transverse axis, and b is half the conjugate axis, we get: $a = 2$, $b = 1$, and $c^2 = a^2 + b^2 \Rightarrow c^2 = 4 + 1 = 5 \Rightarrow c = \sqrt{5}$.

So the focal distance is $\sqrt{5}$, and the foci are $(1 - \sqrt{5}, 0)$ and $(1 + \sqrt{5}, 0)$.
And the vertices are (-1, 0) and (3, 0).

Next, the asymptotes are: $y = \pm(b/a)(x - 1) = \pm\frac{1}{2}(x - 1)$.

Next, assuming e is the eccentricity, we get: $e = c/a = \frac{\sqrt{5}}{2}$.

And next, the directrices are: $x = \pm a/e + 1 = \pm a^2/c + 1 = \pm\frac{4\sqrt{5}}{5} + 1$.

If not quite sure of the idea behind the solution above, follow the steps below:

To begin with, if a hyperbola is centered at **(u, 0)**, and is horizontal, the equation is:
$\dfrac{(x-u)^2}{a^2} - \dfrac{y^2}{b^2} = 1$, where a is half the transverse axis, and b is half the conjugate axis.

And if vertical, its equation is: $\dfrac{y^2}{b^2} - \dfrac{(x-u)^2}{a^2} = 1$, where b is half the transverse axis, and a is half the conjugate axis.

Next, we can put the hyperbola given this way: $\dfrac{(x-1)^2}{2^2} - \dfrac{y^2}{1^2} = 1$.

So the hyperbola given is horizontal, the center is (1, 0), the transverse axis is 4, and the conjugate axis is 2.   What then, about the focal distance?

If c is the focal distance, $2a$ is the transverse axis, and $2b$ is the conjugate axis, we get:
$c^2 = a^2 + b^2$.   So we get: $c^2 = 4 + 1 = 5 \Rightarrow c = \sqrt{5}$.

Next, the center, foci, and vertices are all in the transverse axis. So if the hyperbola is horizontal, the center, foci, and vertices share the same y-coordinate.

The center is the midpoint between the foci, and the focal distance is the distance from the center to each focus, and is c, which is $\sqrt{5}$.

And also, the center is the midpoint between the vertices, too, which are the endpoints of the transverse axis, which is twice the distance from the center to each vertex, and the distance in this case, is a, which is 2.

So since the center is (1, 0), and the hyperbola is horizontal, the foci are $(1-\sqrt{5},0)$ and $(1+\sqrt{5},0)$, and the vertices are (-2 + 1, 0) and (2 + 1, 0), that is, **(-1, 0)** and **(3, 0)**.

Next, a hyperbola has two lines called the asymptotes, the slope of one is b/a, and the other is $-b/a$. And the asymptotes pass through the center, which is (1, 0) in this case. So the asymptotes are: $y = \pm(b/a)(x - 1) \Rightarrow y = \pm\frac{1}{2}(x - 1)$.

Next, the eccentricity of a hyperbola is a ratio, the focal distance over half the transverse axis. So assuming e is the eccentricity, we get: $e = c/a = \frac{\sqrt{5}}{2}$.

And next, a hyperbola has two lines called the directrices, and the distance from each to the center is a ratio, half the transverse over the eccentricity. So since the center is (1, 0), and the hyperbola is horizontal, the directrices are as follows:

$x = \pm a/e + 1 = \pm a^2/c + 1 = \pm\frac{4}{\sqrt{5}} + 1 = \pm\frac{4\sqrt{5}}{5} + 1.$

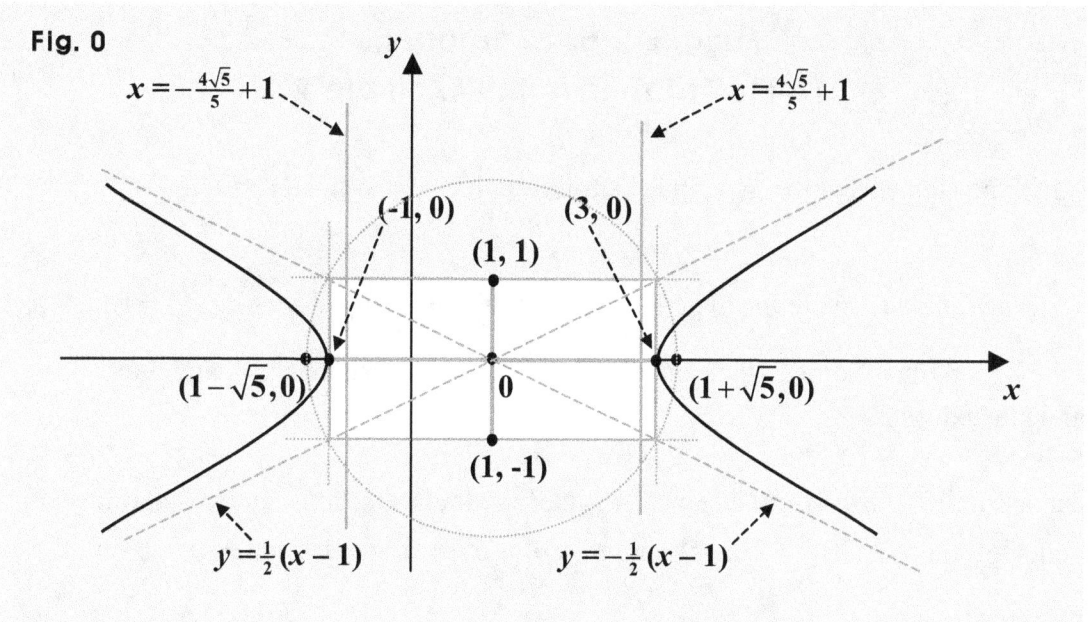

Fig. 0

The hyperbola is: $\dfrac{(x-1)^2}{4} - y^2 = 1,$ often put this way: $\dfrac{(x-1)^2}{2^2} - \dfrac{y^2}{1^2} = 1.$

Note that half the transverse axis is called the semi major axis, too, and half the conjugate axis is called the semi minor axis.

132

Suggestions or Solutions
To the Problem in the Example 2

Find all the elements of the hyperbola as follows: $4(x-2)^2 - (y-1)^2 = -4$.

To begin with, we can put the hyperbola given this way: $\dfrac{(y-1)^2}{2^2} - \dfrac{(x-2)^2}{1^2} = 1$.

So the hyperbola is vertical, the center is (2, 1), the transverse axis is 4, and the conjugate axis is 2.

Next, assuming c is the focal distance, b is half the transverse axis, and a is half the conjugate axis, we get: $b = 2$, $a = 1$, and $c^2 = a^2 + b^2 \Rightarrow c^2 = 1 + 4 = 5 \Rightarrow c = \sqrt{5}$.

So the focal distance is $\sqrt{5}$, and the foci are $(2, 1-\sqrt{5})$ and $(2, 1+\sqrt{5})$.
And the vertices are (2, 3) and (2, -1).

Next, the asymptotes are: $y - 1 = \pm(b/a)(x-2) \Rightarrow y - 1 = \pm 2(x-2)$.

Next, assuming e is the eccentricity, we get: $e = c/b = \frac{\sqrt{5}}{2}$.

And next, the directrices are: $y = \pm b/e + 1 = \pm b^2/c + 1 = \pm \frac{4\sqrt{5}}{5} + 1$.

If not quite sure of the idea behind the solution above, follow the steps below:

To begin with, if a hyperbola is centered at *(u, v)*, and is horizontal, the equation is: $\dfrac{(x-u)^2}{a^2} - \dfrac{(y-v)^2}{b^2} = 1$, where a is half the transverse axis, and b is half the conjugate

axis. And if vertical, its equation is: $\dfrac{(y-v)^2}{b^2} - \dfrac{(x-u)^2}{a^2} = 1$, where b is half the

transverse axis, and a is half the conjugate axis.

Next, we can put the hyperbola given the way below:

$4(x-2)^2 - (y-1)^2 = -4 \Rightarrow \frac{1}{4}(y-1)^2 - (x-2)^2 = 1 \Rightarrow \dfrac{(y-1)^2}{2^2} - \dfrac{(x-2)^2}{1^2} = 1.$

So the hyperbola given is vertical, the center is (2, 1), the transverse axis is 4, and the conjugate axis is 2.   What then, about the focal distance?

If c is the focal distance, $2b$ is the transverse axis, and $2a$ is the conjugate axis, we get: $c^2 = a^2 + b^2.$ So we get: $c^2 = 1 + 4 = 5 \Rightarrow c = \sqrt{5}.$

Next, the center, foci, and vertices are all in the transverse axis. So if the hyperbola is vertical, the center, foci, and vertices share the same x-coordinate.

The center is the midpoint between the foci, and the focal distance is the distance from the center to each focus, and is c, which is $\sqrt{5}.$

And also, the center is the midpoint between the vertices, too, which are the endpoints of the transverse axis, which is twice the distance from the center to each vertex, and the distance in this case, is b, which is 2.

So since the center is (2, 1), and the hyperbola is vertical, the foci are $(0 + 2, 1 - \sqrt{5})$ and $(0 + 2, 1 + \sqrt{5}),$ that is, $(2, 1 - \sqrt{5})$ and $(2, 1 + \sqrt{5}),$ and the vertices are $(0 + 2, -2 + 1)$ and $(0 + 2, 2 + 1),$ that is, **(2, -1)** and **(2, 3)**.

Next, a hyperbola has two lines called the asymptotes, the slope of one is b/a, and the other is $-b/a$. And the asymptotes pass through the center, which is (2, 1) in this case. So the asymptotes are: $y - 1 = \pm(b/a)(x - 2) \Rightarrow y - 1 = \pm 2(x - 2).$

Next, the eccentricity of a hyperbola is a ratio, the focal distance over half the transverse axis. So assuming e is the eccentricity, we get: $e = c/b = \frac{\sqrt{5}}{2}.$

And next, a hyperbola has two lines called the directrices, and the distance from each to the center is a ratio, half the transverse over the eccentricity. So since the center is (2, 1), and the hyperbola is vertical, the directrices are as follows:

$$y = \pm b/e + 1 = \pm b^2/c + 1 = \pm \frac{4}{\sqrt{5}} + 1 = \pm \frac{4\sqrt{5}}{5} + 1.$$

Fig. 0

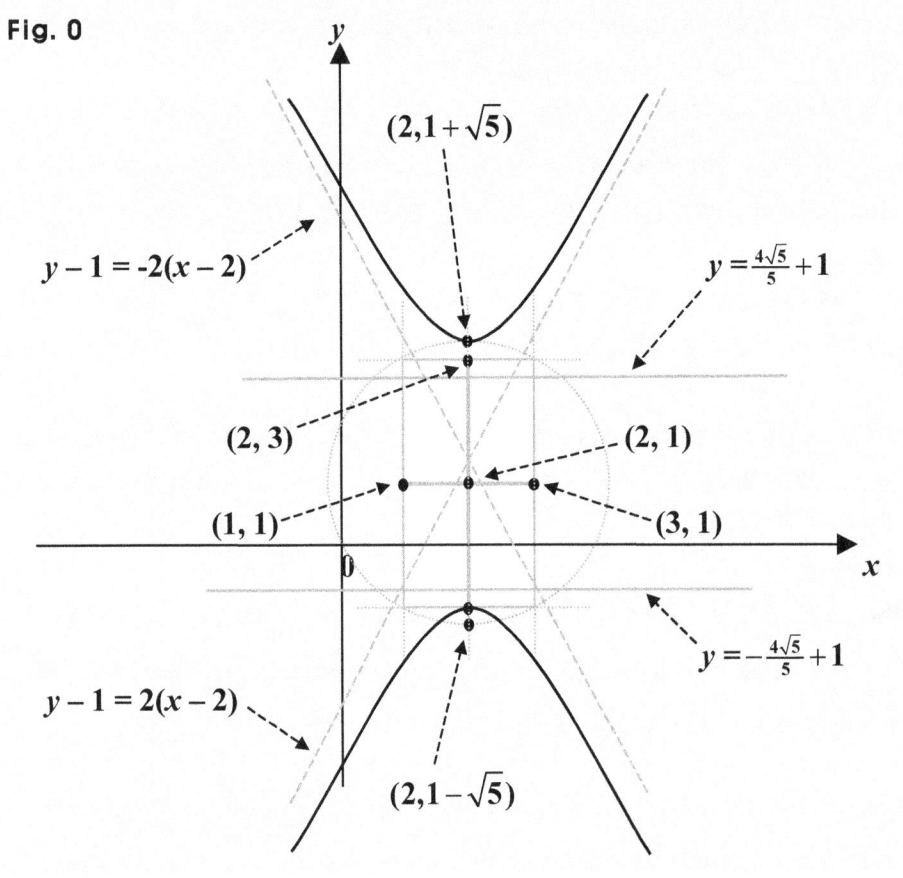

The hyperbola is: $\dfrac{(y-1)^2}{4}-(x-2)^2=1,$ often put this way: $\dfrac{(y-1)^2}{2^2}-\dfrac{(x-2)^2}{1^2}=1.$

Note that half the transverse axis is called the semi major axis, too, and half the conjugate axis is called the semi minor axis.

Suggestions or Solutions
To the Problem in the Example 3

Find all the elements of the hyperbola as follows: $1 + 16(y-1)^2 - 25(x-2)^2 = 0$.

To begin with, we can put the hyperbola given this way: $\dfrac{(x-2)^2}{(\frac{1}{5})^2} - \dfrac{(y-1)^2}{(\frac{1}{4})^2} = 1$.

So the hyperbola is horizontal, the center is $(2, 1)$, the transverse axis is $\frac{2}{5}$, and the conjugate axis is $\frac{1}{2}$.

Next, assuming c is the focal distance, a is half the transverse axis, and b is half the conjugate axis, we get: $a = \frac{1}{5}, b = \frac{1}{4}$, and $c^2 = a^2 + b^2 \Rightarrow c^2 = (\frac{1}{5})^2 + (\frac{1}{4})^2 = \frac{41}{400} \Rightarrow c = \frac{\sqrt{41}}{20}$.

So the focal distance is $\frac{\sqrt{41}}{20}$, and the foci are $(2 + \frac{\sqrt{41}}{20}, 1)$ and $(2 - \frac{\sqrt{41}}{20}, 1)$.

And the vertices are $(\frac{11}{5}, 1)$ and $(\frac{9}{5}, 1)$.

Next, the asymptotes are: $y - 2 = \pm(b/a)(x-1) \Rightarrow y - 2 = \pm\frac{5}{4}(x-1)$.

Next, assuming e is the eccentricity, we get: $e = c/a = \frac{\sqrt{41}}{4}$.

And next, the directrices are: $x = \pm a/e + 2 = \pm a^2/c + 2 = \pm\frac{4\sqrt{41}}{205} + 2$.

If not quite sure of the idea behind the solution above, follow the steps below:

To begin with, if a hyperbola is centered at (u, v), and is horizontal, the equation is: $\dfrac{(x-u)^2}{a^2} - \dfrac{(y-v)^2}{b^2} = 1$, where a is half the transverse axis, and b is half the conjugate axis. And if vertical, its equation is: $\dfrac{(y-v)^2}{b^2} - \dfrac{(x-u)^2}{a^2} = 1$, where b is half the transverse axis, and a is half the conjugate axis.

Next, we can put the hyperbola given the way below:

$$1 + 16(y-1)^2 - 25(x-2)^2 = 0 \Rightarrow 25(x-2)^2 - 16(y-1)^2 = 1 \Rightarrow \dfrac{(x-2)^2}{(\frac{1}{5})^2} - \dfrac{(y-1)^2}{(\frac{1}{4})^2} = 1.$$

So the hyperbola is horizontal, the center is (2, 1), the transverse axis is $\frac{2}{5}$, and the conjugate axis is $\frac{1}{2}$.   What then, about the focal distance?

If c is the focal distance, $2a$ is the transverse axis, and $2b$ is the conjugate axis, we get: $c^2 = a^2 + b^2$.   So we get: $c^2 = (\frac{1}{5})^2 + (\frac{1}{4})^2 = \frac{1}{25} + \frac{1}{16} = \frac{41}{400} \Rightarrow c = \frac{\sqrt{41}}{20}$.

Next, the center, foci, and vertices are all in the transverse axis. So if the hyperbola is horizontal, the center, foci, and vertices share the <u>same y-coordinate</u>.

The center is the midpoint between the foci, and the focal distance is the distance from the center to each focus, and is c, which is $\frac{\sqrt{41}}{20}$.

And also, the center is the midpoint between the <u>vertices</u>, too, which are the <u>endpoints</u> of the <u>transverse axis</u>, which is twice the distance from the center to each vertex, and the distance in this case, is a, which is $\frac{1}{5}$.

So since the center is (2, 1), and the hyperbola is horizontal, the foci are $(\frac{\sqrt{41}}{20} + 2, 0 + 1)$ and $(2 - \frac{\sqrt{41}}{20}, 0 + 1)$, that is, $(\frac{\sqrt{41}}{20} + 2, 1)$ and $(2 - \frac{\sqrt{41}}{20}, 1)$, and the vertices are $(\frac{1}{5} + 2, 0 + 1)$ and $(2 - \frac{1}{5}, 0 + 1)$, that is, $(\frac{11}{5}, 1)$ and $(\frac{9}{5}, 1)$.

Next, a hyperbola has two lines called the <u>asymptotes</u>, the slope of one is b/a, and the other is $-b/a$. And the asymptotes pass through the center, which is (1, 2) in this case. So the asymptotes are: $y - 2 = \pm(b/a)(x - 1) \Rightarrow y - 2 = \pm\frac{5}{4}(x - 1)$.

Next, the <u>eccentricity</u> of a hyperbola is a ratio, <u>the focal distance over half the transverse axis</u>. So assuming e is the eccentricity, we get: $e = c/a = \frac{\sqrt{41}}{20} \cdot 5 = \frac{\sqrt{41}}{4}$.

And next, a hyperbola has two lines called the <u>directrices</u>, and the distance from each to the center is a ratio, <u>half the transverse over the eccentricity</u>. So since the center is (2, 1), and the hyperbola is <u>horizontal</u>, the directrices are as follows:

$$x = \pm a/e + 2 = \pm a^2/c + 2 = \pm \frac{20}{25\sqrt{41}} + 2 = \pm \frac{4}{5\sqrt{41}} + 2 = \pm \frac{4\sqrt{41}}{5 \cdot 41} + 2 = \pm \frac{4\sqrt{41}}{205} + 2.$$

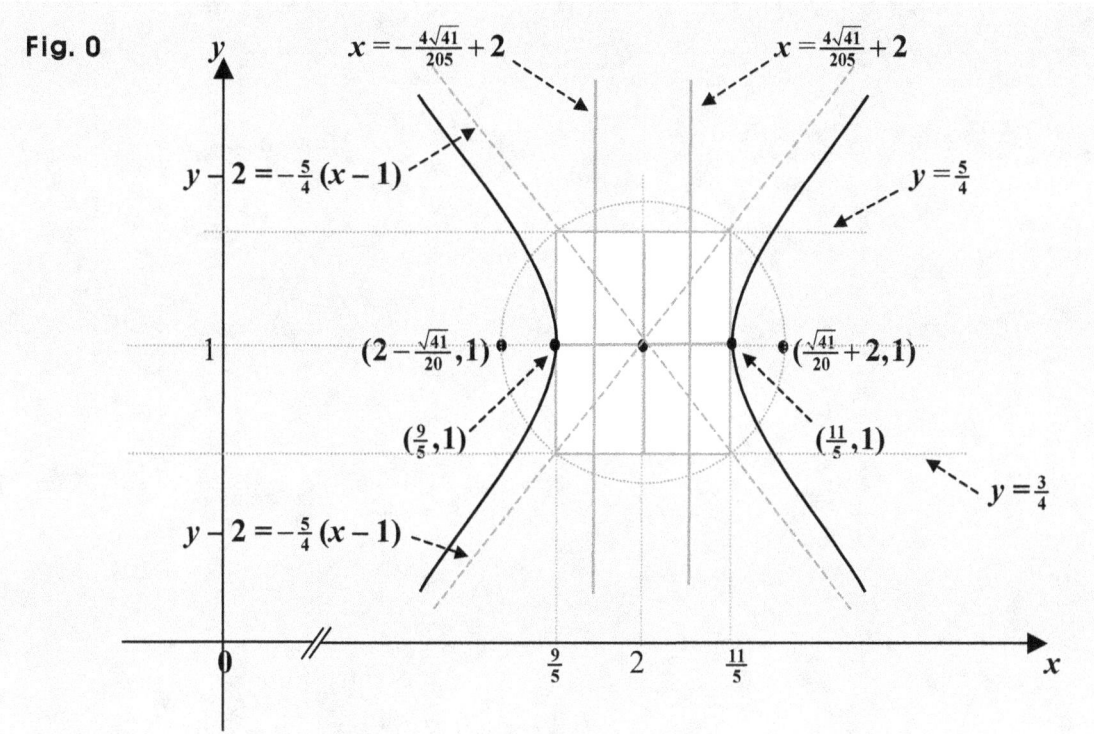

Fig. 0

The hyperbola is: $25(x-2)^2 - 16(y-1)^2 = 1$, often put this way: $\dfrac{(x-2)^2}{\left(\frac{1}{5}\right)^2} - \dfrac{(y-1)^2}{\left(\frac{1}{4}\right)^2} = 1$.

Note that half the transverse axis is called the semi major axis, too, and half the conjugate axis is called the semi minor axis.

Examples 5 in Hyperbolas

Find if each hyperbola below is horizontal or vertical, the center, foci, vertices, transverse axis, conjugate axis, asymptotes, eccentricity, and directrices.

0. $4x^2 + 8x + 4 - y^2 + 1 = 0$

1. $1 - x^2 + 8y + 4 + 4y^2 = 0$

2. $16x^2 + 32x - 25y^2 - 100y - 484 = 0$

3. $12x - 3x^2 + 8y + 4y^2 + 4 = 0$

Suggestions or Solutions
To the Problem in the Example 0

Find all the elements of the hyperbola as follows: $4x^2 + 8x + 4 - y^2 + 1 = 0$.

To begin with, we can put the hyperbola given this way: $\dfrac{y^2}{1^2} - \dfrac{(x+1)^2}{(\frac{1}{2})^2} = 1$.

So the hyperbola is vertical, the center is (-1, 0), the transverse axis is 2, and the conjugate axis is 1.

Next, assuming c is the focal distance, b is half the transverse axis, and a is half the conjugate axis, we get: $b = 1$, $a = \frac{1}{2}$, and $c^2 = a^2 + b^2 \Rightarrow c^2 = \frac{1}{4} + 1 = \frac{5}{4} \Rightarrow c = \frac{\sqrt{5}}{2}$.

So the focal distance is $\frac{\sqrt{5}}{2}$, and the foci are $(-1, \frac{\sqrt{5}}{2})$ and $(-1, -\frac{\sqrt{5}}{2})$

And the vertices are (-1, 1) and (-1, -1).

Next, the asymptotes are: $y = \pm(b/a)(x + 1) \Rightarrow y = \pm 2(x + 1)$.

Next, assuming e is the eccentricity, we get: $e = c/b = \frac{\sqrt{5}}{2}$.

And next, the directrices are: $y = \pm b/e = \pm b^2/c = \pm \frac{2\sqrt{5}}{5}$.

If not quite sure of the idea behind the solution above, follow the steps below:

To begin with, if a hyperbola is centered at (u, v), and is horizontal, the equation is: $\dfrac{(x-u)^2}{a^2} - \dfrac{(y-v)^2}{b^2} = 1$, where a is half the transverse axis, and b is half the conjugate axis. And if vertical, its equation is: $\dfrac{(y-v)^2}{b^2} - \dfrac{(x-u)^2}{a^2} = 1$, where b is half the transverse axis, and a is half the conjugate axis.

Next, we can put the hyperbola given the way below:

$4x^2 + 8x + 4 - y^2 + 1 = 0 \Rightarrow 4(x^2 + 2x + 1) - y^2 + 1 = 4(x + 1)^2 - y^2 + 1 = 0$

$\Rightarrow y^2 - 4(x + 1)^2 = 1 \Rightarrow \dfrac{y^2}{1^2} - \dfrac{(x+1)^2}{(\frac{1}{2})^2} = 1$.

So the hyperbola given is vertical, the center is (-1, 0), the transverse axis is 2, and the conjugate axis is 1.   What then, about the focal distance?

If c is the focal distance, **2b** is the transverse axis, and **2a** is the conjugate axis, we get: $c^2 = a^2 + b^2$.   So we get: $c^2 = \frac{1}{4} + 1 = \frac{5}{4} \Rightarrow c = \frac{\sqrt{5}}{2}$.

Next, the center, foci, and vertices are all in the transverse axis. So if the hyperbola is <u>vertical</u>, the center, foci, and vertices share the <u>same x-coordinate</u>.

The center is the midpoint between the foci, and the focal distance is the distance from the center to each focus, and is c, which is $\frac{\sqrt{5}}{2}$.

And also, the center is the midpoint between the <u>vertices</u>, too, which are the <u>endpoints</u> of the <u>transverse axis</u>, which is twice the distance from the center to each vertex, and the distance in this case, is b, which is 1.

So since the center is (-1, 0), and the hyperbola is vertical, the foci are $\left(0 - 1, \frac{\sqrt{5}}{2}\right)$ and $\left(0 - 1, -\frac{\sqrt{5}}{2}\right)$ that is, $\left(-1, \frac{\sqrt{5}}{2}\right)$ and $\left(-1, -\frac{\sqrt{5}}{2}\right)$ and the vertices are $(0 - 1, 1)$ and $(0 - 1, -1)$, that is, **(-1, 1)** and **(-1, -1)**.

Next, a hyperbola has two lines called the <u>asymptotes</u>, the slope of one is **b/a**, and the other is **−b/a**. And the asymptotes pass through the center, which is (-1, 0) in this case. So the asymptotes are: $y = \pm(b/a)(x + 1) \Rightarrow y = \pm 2(x + 1)$.

Next, the <u>eccentricity</u> of a hyperbola is a ratio, <u>the focal distance over half the transverse axis</u>. So assuming e is the eccentricity, we get: $e = c/b = \frac{\sqrt{5}}{2}$.

And next, a hyperbola has two lines called the <u>directrices</u>, and the distance from each to the center is a ratio, <u>half the transverse over the eccentricity</u>. So since the center is (-1, 0), and the hyperbola is <u>vertical</u>, the directrices are as follows:

$$y = \pm b/e = \pm b^2/c = \pm \frac{2}{\sqrt{5}} = \pm \frac{2\sqrt{5}}{5}.$$

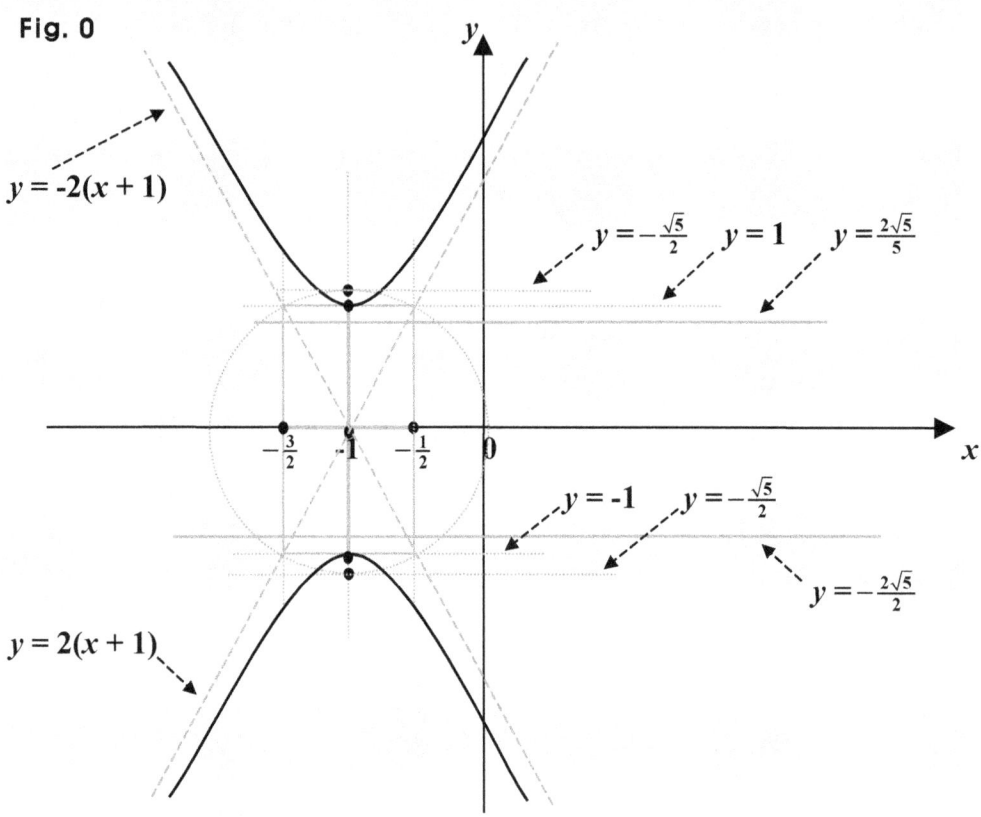

Fig. 0

The hyperbola is: $y^2 - 4(x + 1)^2 = 1$, which is often put this way: $\dfrac{y^2}{1^2} - \dfrac{(x+1)^2}{\left(\frac{1}{2}\right)^2} = 1$.

Note that half the transverse axis is called the semi major axis, too, and half the conjugate axis is called the semi minor axis.

Suggestions or Solutions
To the **Problem** in the Example 1

Find all the elements of the hyperbola as follows: $1 - x^2 + 8y + 4 + 4y^2 = 0$.

To begin with, we can put the hyperbola given this way: $\dfrac{x^2}{1^2} - \dfrac{(y+1)^2}{(\frac{1}{2})^2} = 1$.

So the hyperbola is horizontal, the center is (0, -1), the transverse axis is 2, and the conjugate axis is 1.

Next, assuming c is the focal distance, a is half the transverse axis, and b is half the conjugate axis, we get: $a = 1$, $b = \frac{1}{2}$, and $c^2 = a^2 + b^2 \Rightarrow c^2 = 1 + \frac{1}{4} = \frac{5}{4} \Rightarrow c = \frac{\sqrt{5}}{2}$.

So the focal distance is $\frac{\sqrt{5}}{2}$, and the foci are $(\frac{\sqrt{5}}{2}, -1)$ and $(-\frac{\sqrt{5}}{2}, -1)$.
And the vertices are (-1, -1) and (1, -1).

Next, the asymptotes are: $y + 1 = \pm(b/a)x \Rightarrow y + 1 = \pm\frac{1}{2}x$.

Next, assuming e is the eccentricity, we get: $e = c/a = \frac{\sqrt{5}}{2}$.

And next, the directrices are: $x = \pm a/e = \pm a^2/c = \pm\frac{2\sqrt{5}}{5}$.

If not quite sure of the idea behind the solution above, follow the steps below:

To begin with, if a hyperbola is centered at (u, v), and is horizontal, the equation is: $\dfrac{(x-u)^2}{a^2} - \dfrac{(y-v)^2}{b^2} = 1$, where a is half the transverse axis, and b is half the conjugate axis. And if vertical, its equation is: $\dfrac{(y-v)^2}{b^2} - \dfrac{(x-u)^2}{a^2} = 1$, where b is half the transverse axis, and a is half the conjugate axis.

Next, we can put the hyperbola given the way below:

$1 - x^2 + 8y + 4 + 4y^2 = 0 \Rightarrow 4(y^2 + 2y + 1) - x^2 + 1 = 4(y+1)^2 - x^2 + 1 = 0$

$\Rightarrow x^2 - 4(y+1)^2 = 1 \Rightarrow \dfrac{x^2}{1^2} - \dfrac{(y+1)^2}{(\frac{1}{2})^2} = 1$.

So the hyperbola given is horizontal, the center is (0, -1), the transverse axis is 2, and the conjugate axis is 1.   What then, about the focal distance?

If *c* is the focal distance, **2a** is the transverse axis, and **2b** is the conjugate axis, we get: $c^2 = a^2 + b^2$.   So we get: $c^2 = 1 + \frac{1}{4} = \frac{5}{4} \Rightarrow c = \frac{\sqrt{5}}{2}$.

Next, the center, foci, and vertices are all in the transverse axis. So if the hyperbola is horizontal, the center, foci, and vertices share the same *y*-coordinate.

The center is the midpoint between the foci, and the focal distance is the distance from the center to each focus, and is *c*, which is $\frac{\sqrt{5}}{2}$.

And also, the center is the midpoint between the vertices, too, which are the endpoints of the transverse axis, which is twice the distance from the center to each vertex, and the distance in this case, is *a*, which is 1.

So since the center is (0, -1), and the hyperbola is vertical, the foci are $(\frac{\sqrt{5}}{2}, 0-1)$ and $(-\frac{\sqrt{5}}{2}, 0-1)$, that is, $(\frac{\sqrt{5}}{2}, -1)$ and $(-\frac{\sqrt{5}}{2}, -1)$, and the vertices are (1, 0 − 1) and (-1, 0 − 1), that is, **(1, -1)** and **(-1, -1)**.

Next, a hyperbola has two lines called the asymptotes, the slope of one is **b/a**, and the other is **−b/a**. And the asymptotes pass through the center, which is (0, -1) in this case. So the asymptotes are: $y + 1 = \pm(b/a)x \Rightarrow y + 1 = \pm\frac{1}{2}x$.

Next, the eccentricity of a hyperbola is a ratio, the focal distance over half the transverse axis. So assuming *e* is the eccentricity, we get: $e = c/a = \frac{\sqrt{5}}{2}$.

And next, a hyperbola has two lines called the directrices, and the distance from each to the center is a ratio, half the transverse over the eccentricity. So since the center is (0, -1), and the hyperbola is horizontal, the directrices are as follows:

$$x = \pm a/e = \pm a^2/c = \pm \frac{2}{\sqrt{5}} = \pm \frac{2\sqrt{5}}{5}.$$

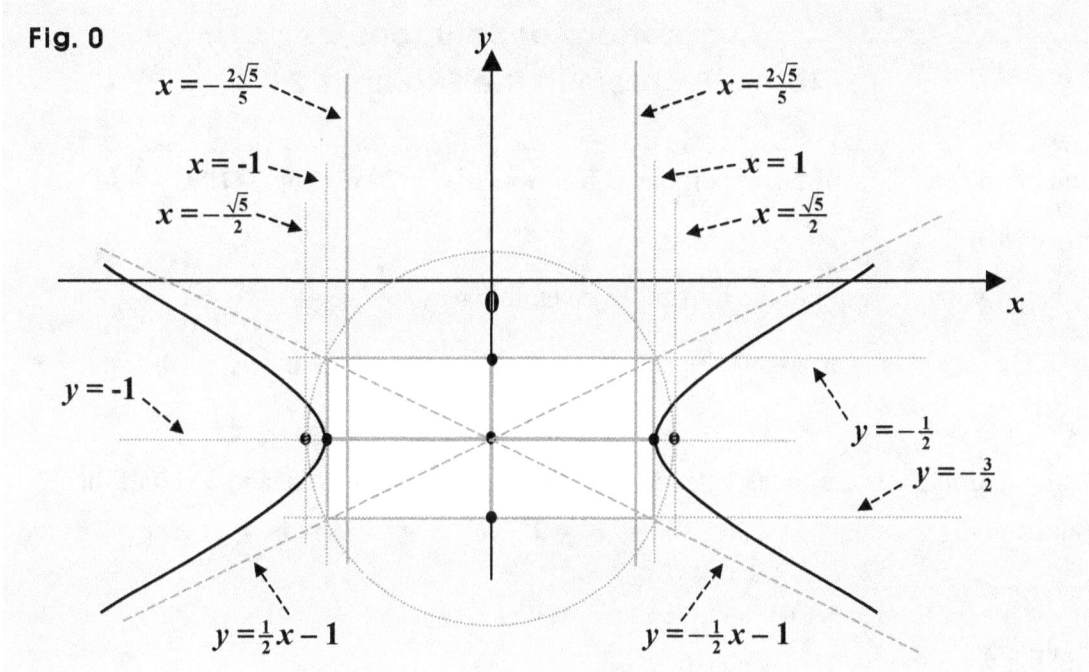

Fig. 0

The hyperbola is: $x^2 - 4(y + 1)^2 = 1$, which is often put this way: $\dfrac{x^2}{1^2} - \dfrac{(y+1)^2}{(\frac{1}{2})^2} = 1$.

Note that half the transverse axis is called the semi major axis, too, and half the conjugate axis is called the semi minor axis.

Suggestions or Solutions
To the Problem in the Example 2

Find all the elements of the hyperbola as follows: $16x^2 + 32x - 25y^2 - 100y - 484 = 0$.

To begin with, we can put the hyperbola given this way: $\dfrac{(x+1)^2}{5^2} - \dfrac{(y+2)^2}{4^2} = 1$.

So the hyperbola is horizontal, the center is (-1, -2), the transverse axis is 10, and the conjugate axis is 8.

Next, assuming c is the focal distance, a is half the transverse axis, and b is half the conjugate axis, we get: $a = 5$, $b = 4$, and $c^2 = a^2 + b^2 \Rightarrow c^2 = 25 + 16 = 41 \Rightarrow c = \sqrt{41}$.

So the focal distance is $\sqrt{41}$, and the foci are $(\sqrt{41} - 1, -2)$ and $(-\sqrt{41} - 1, -2)$.
And the vertices are (4, -2) and (-6, -2).

Next, the asymptotes are: $y + 2 = \pm(b/a)(x + 1) \Rightarrow y + 2 = \pm\frac{1}{2}(x + 1)$.

Next, assuming e is the eccentricity, we get: $e = c/a = \frac{\sqrt{41}}{5}$.

And next, the directrices are: $x = \pm a/e - 1 = \pm a^2/c - 1 = \pm \frac{25\sqrt{41}}{41} - 1$.

If not quite sure of the idea behind the solution above, follow the steps below:

To begin with, if a hyperbola is centered at (u, v), and is horizontal, the equation is:
$\dfrac{(x-u)^2}{a^2} - \dfrac{(y-v)^2}{b^2} = 1$, where a is half the transverse axis, and b is half the conjugate

axis. And if vertical, its equation is: $\dfrac{(y-v)^2}{b^2} - \dfrac{(x-u)^2}{a^2} = 1$, where b is half the

transverse axis, and a is half the conjugate axis.

Next, we can put the hyperbola given the way below:

$16x^2 + 32x - 25y^2 - 100y - 484 = 16(x^2 + 2x + 1 - 1) - 25(y^2 + 4y + 4 - 4) - 484$

$= 16(x + 1)^2 - 16 - 25(y + 2)^2 + 100 - 484 = 16(x + 1)^2 - 25(y + 2)^2 - 400 = 0 \Rightarrow$

$\dfrac{16}{400}(x+1)^2 - \dfrac{25}{400}(y+2)^2 - 1 = 0 \Rightarrow \dfrac{1}{25}(x+1)^2 - \dfrac{1}{16}(y+2)^2 = 1 \Rightarrow \dfrac{(x+1)^2}{5^2} - \dfrac{(y+2)^2}{4^2} = 1$.

So the hyperbola given is horizontal, the center is (-1, -2), the transverse axis is 10, and the conjugate axis is 8. What then, about the focal distance?

If c is the focal distance, $2a$ is the transverse axis, and $2b$ is the conjugate axis, we get: $c^2 = a^2 + b^2$. So we get: $c^2 = 25 + 15 = 41 \Rightarrow c = \sqrt{41}$.

Next, the center, foci, and vertices are all in the transverse axis. So if the hyperbola is horizontal, the center, foci, and vertices share the <u>same y-coordinate</u>.

The center is the midpoint between the foci, and the focal distance is the distance from the center to each focus, and is c, which is $\sqrt{41}$.

And also, the center is the midpoint between the <u>vertices</u>, too, which are the <u>endpoints</u> of the <u>transverse axis</u>, which is twice the distance from the center to each vertex, and the distance in this case, is a, which is 5.

So since the center is (-1, -2), and the hyperbola is vertical, the foci are $(\sqrt{41} - 1, 0 - 2)$ and $(-\sqrt{41} - 1, 0 - 2)$ that is, $(\sqrt{41} - 1, -2)$ and $(-\sqrt{41} - 1, -2)$, and next, the two vertices are (5 − 1, 0 − 2) and (-5 − 1, 0 − 2), that is, **(4, -2)** and **(-6, -2)**.

Next, a hyperbola has two lines called the <u>asymptotes</u>, the slope of one is b/a, and the other is $-b/a$. And the asymptotes pass through the center, which is (-1, -2) in this case. So the asymptotes are: $y + 2 = \pm(b/a)(x + 1) \Rightarrow y + 2 = \pm\frac{1}{2}(x + 1)$.

Next, the <u>eccentricity</u> of a hyperbola is a ratio, <u>the focal distance over half the transverse axis</u>. So assuming e is the eccentricity, we get: $e = c/a = \frac{\sqrt{41}}{5}$.

And next, a hyperbola has two lines called the <u>directrices</u>, and the distance from each to the center is a ratio, <u>half the transverse over the eccentricity</u>.
So since the center is (-1, -2), and the hyperbola is <u>horizontal</u>, the directrices are as follows: $x = \pm a/e - 1 = \pm a^2/c - 1 = \pm\frac{25}{\sqrt{41}} - 1 = \pm\frac{25\sqrt{41}}{41} - 1$.

Fig. 0

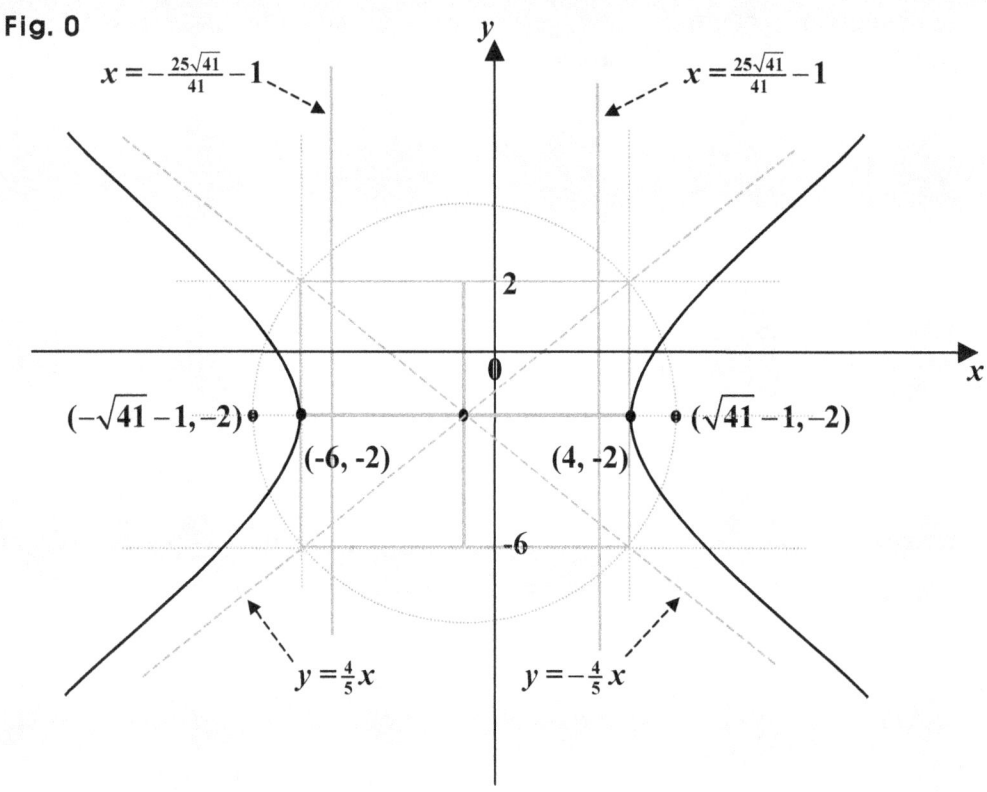

The hyperbola is: $\dfrac{(x+1)^2}{25} - \dfrac{(y+2)^2}{16} = 1,$ often put this way: $\dfrac{(x+1)^2}{5^2} - \dfrac{(y+2)^2}{4^2} = 1.$

Note that half the transverse axis is called the semi major axis, too, and half the conjugate axis is called the semi minor axis.

Suggestions or Solutions
To the Problem in the Example 3

Find all the elements of the hyperbola as follows: $12x - 3x^2 + 8y + 4y^2 + 4 = 0$.

To begin with, we can put the hyperbola given this way: $\dfrac{(x-2)^2}{2^2} - \dfrac{(y+1)^2}{(\sqrt{3})^2} = 1$.

So the hyperbola is horizontal, the center is (2, -1), the transverse axis is 4, and the conjugate axis is $2\sqrt{3}$.

Next, assuming c is the focal distance, a is half the transverse axis, and b is half the conjugate axis, we get: $a = 5$, $b = \sqrt{3}$, and $c^2 = a^2 + b^2 \Rightarrow c^2 = 25 + 3 = 28 \Rightarrow c = \sqrt{28}$.

So the focal distance is $\sqrt{28}$, and the foci are $(\sqrt{28} + 2, -1)$ and $(-\sqrt{28} + 2, -1)$.
And the vertices are (0, -1) and (4, -1).

Next, the asymptotes are: $y + 1 = \pm(b/a)(x - 2) \Rightarrow y + 1 = \pm\frac{1}{2}(x - 2)$.

Next, assuming e is the eccentricity, we get: $e = c/a = \sqrt{7}$.

And next, the directrices are: $x = \pm a/e + 2 = \pm a^2/c + 2 = \pm\frac{2\sqrt{7}}{7} + 2$.

If not quite sure of the idea behind the solution above, follow the steps below:

To begin with, if a hyperbola is centered at (u, v), and is horizontal, the equation is: $\dfrac{(x-u)^2}{a^2} - \dfrac{(y-v)^2}{b^2} = 1$, where a is half the transverse axis, and b is half the conjugate axis.

Next, we can put the hyperbola given the way below:

$12x - 3x^2 + 8y + 4y^2 + 4 = 0 \Rightarrow 3x^2 - 12x - 4y^2 - 8y - 4 = 0$

$\Rightarrow 3(x^2 - 4x + 4 - 4) - 4(y^2 + 2y + 1) = 3(x-2)^2 - 4(y+1)^2 - 12 = 0$

$\Rightarrow \dfrac{(x-2)^2}{4} - \dfrac{(y+1)^2}{3} = 1 \Rightarrow \dfrac{(x-2)^2}{2^2} - \dfrac{(y+1)^2}{(\sqrt{3})^2} = 1$.

So the hyperbola given is horizontal, the center is (2, -1), the transverse axis is 4, and the conjugate axis is $2\sqrt{3}$.   What then, about the focal distance?

If c is the focal distance, $2a$ is the transverse axis, and $2b$ is the conjugate axis, we get: $c^2 = a^2 + b^2$.   So we get: $c^2 = 25 + 3 = 28 \Rightarrow c = \sqrt{28}$.

Next, the center, foci, and vertices are all in the transverse axis. So if the hyperbola is <u>horizontal</u>, the center, foci, and vertices share the <u>same y-coordinate</u>.

The center is the midpoint between the foci, and the focal distance is the distance from the center to each focus, and is c, which is $\sqrt{28}$.

And also, the center is the midpoint between the <u>vertices</u>, too, which are the <u>endpoints</u> of the <u>transverse axis</u>, which is twice the distance from the center to each vertex, and the distance in this case, is a, which is 2.

So since the center is (2, -1), and the hyperbola is vertical, the foci are $(\sqrt{28} + 2, 0 - 1)$ and $(-\sqrt{28} + 2, 0 - 1)$, that is, $(\sqrt{28} + 2, -1)$ and $(-\sqrt{28} + 2, -1)$, and the vertices are (-2 + 2, 0 – 1) and (2 + 2, 0 – 1), that is, **(0, -1)** and **(4, -1)**.

Next, a hyperbola has two lines called the <u>asymptotes</u>, the slope of one is b/a, and the other is $-b/a$. And the asymptotes pass through the center, which is (2, -1) in this case. So the asymptotes are: $y + 1 = \pm(b/a)(x - 2) \Rightarrow y + 1 = \pm\frac{1}{2}(x - 2)$.

Next, the <u>eccentricity</u> of a hyperbola is a ratio, <u>the focal distance over half the transverse axis</u>. So assuming e is the eccentricity, we get: $e = c/a = \frac{\sqrt{28}}{2} = \frac{2\sqrt{7}}{2} = \sqrt{7}$.

And next, a hyperbola has two lines called the <u>directrices</u>, and the distance from each to the center is a ratio, <u>half the transverse over the eccentricity</u>. So since the center is (2, -1), and the hyperbola is <u>horizontal</u>, the directrices are as follows:

$x = \pm a/e + 2 = \pm a^2/c + 2 = \pm\frac{4}{\sqrt{28}} + 2 = \pm\frac{4}{2\sqrt{7}} + 2 = \pm\frac{2\sqrt{7}}{7} + 2.$

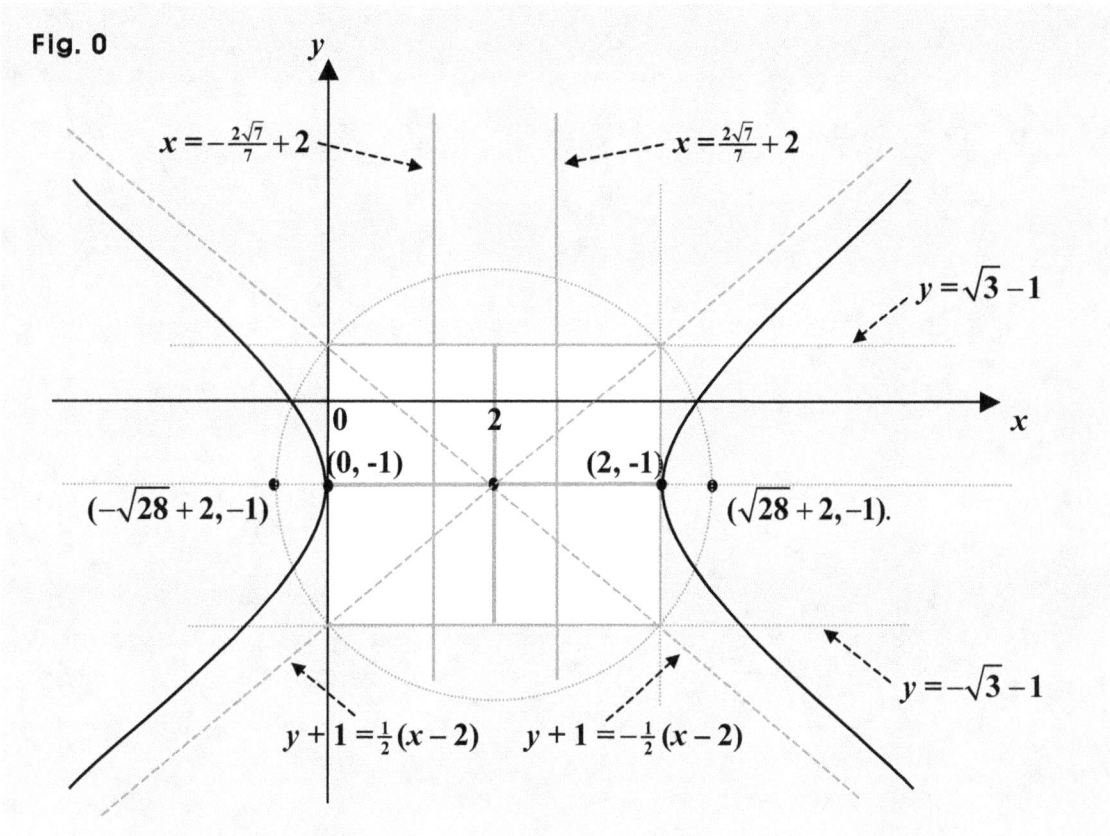

Fig. 0

The hyperbola is: $\dfrac{(x-2)^2}{4} - \dfrac{(y+1)^2}{3^2} = 1$, often put this way: $\dfrac{(x-2)^2}{2^2} - \dfrac{(y+1)^2}{(\sqrt{3})^2} = 1$.

Note that half the transverse axis is called the semi major axis, too, and half the conjugate axis is called the semi minor axis.

Examples 6 in Hyperbolas

In each example below, the hyperbola is in the *x-y* plane.

0. Find the hyperbola where the foci and vertices are: (-1, 3), (4, 3), (-2, 3), and (3, 3).

1. Find the hyperbola that has a point at (-3, 9/4) and has the foci at (-3, 0) and (7, 0).

2. Find the hyperbola that has a point at $(1, 3 - 2\sqrt{2})$, and has the vertices at (4, 5) and (4, 1).

3. Find the hyperbola of which the eccentricity is 5/3 and a directrix is a line $x = 14/5$, which is corresponding a focus at (6, 2).

Suggestions or Solutions
To the Problem in the Example 0

Find the hyperbola where the foci and vertices are: (-1, 3), (4, 3), (-2, 3), and (3, 3).

To begin with, the hyperbola is horizontal, the vertices are (-1, 3) and (3, 3), the foci are (4, 3) and (-2, 3), and the center is (1, 3).

So next, assuming a is half the transverse axis, and c is the focal distance, we get: $a = 2$, and $c = 3$.

Thus, next, assuming b is half the conjugate axis, we get:
$c^2 = a^2 + b^2 \Rightarrow b^2 = c^2 - a^2 = 9 - 4 = 5 \Rightarrow b = \sqrt{5}$.

So the hyperbola is: $\dfrac{(x-1)^2}{4} + \dfrac{(y-3)^2}{5} = 1$, often put this way: $\dfrac{(x-1)^2}{2^2} - \dfrac{(y-3)^2}{(\sqrt{5})^2} = 1$.

If not quite sure of the idea behind the processes above, follow the steps below:

To begin with, if a hyperbola is centered at (u, v), and is horizontal, the equation is:
$\dfrac{(x-u)^2}{a^2} - \dfrac{(y-v)^2}{b^2} = 1$, where a is half the transverse axis, and b is half the conjugate axis. If vertical though, its equation is: $\dfrac{(y-v)^2}{b^2} - \dfrac{(x-u)^2}{a^2} = 1$, where b is half the transverse axis, and a is half the conjugate axis.

And if <u>horizontal</u>, the center, foci, and vertices share the <u>same y-coordinate</u>.

Next, we can notice that all the points given have the <u>same y-coordinate</u>.
So the hyperbola we want to find is horizontal.

Next, the vertices are between the foci.
So the vertices are (-1, 3) and (3, 3), and the foci are (4, 3) and (-2, 3).
And the transverse axis is the distance between the vertices, and thus, is: 4.

What then, about the conjugate axis?

Assuming c is the focal distance, $2a$ is the transverse axis, $2b$ is the conjugate axis, we get: $c^2 = a^2 + b^2$. And we know: $2a = 4$. So we get: $a = 2$, and $b^2 = c^2 - a^2 = c^2 - 4$. What then, about c?

The focal distance is the distance from the center to a focus, which is in this case, (4, 3) or (-2, 3). So we want to get the center.

And the center is the midpoint between the foci.
So the center is: {(-2 + 4)/2, (3 + 3)/2}, and thus, is: (1, 3). So the focal distance c is 3.

And thus, we get: $b^2 = c^2 - 4 = 9 - 4 = 5 \Rightarrow b = \sqrt{5}$.

So the hyperbola is: $\dfrac{(x-1)^2}{4} + \dfrac{(y-3)^2}{5} = 1$, often put this way: $\dfrac{(x-1)^2}{2^2} - \dfrac{(y-3)^2}{(\sqrt{5})^2} = 1$.

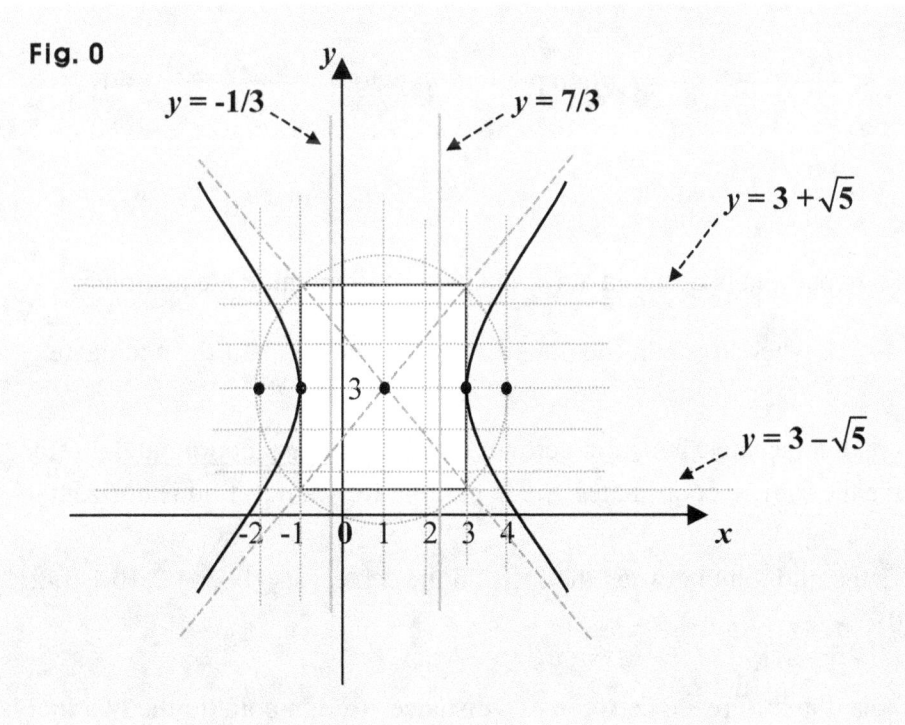

Fig. 0

$y = -1/3$

$y = 7/3$

$y = 3 + \sqrt{5}$

$y = 3 - \sqrt{5}$

Suggestions or Solutions
To the Problem in the Example 1

Find the hyperbola that has a point at (-3, 9/4) and has the foci at (-3, 0) and (7, 0).

To begin with, the hyperbola is horizontal, and the center is (2, 0).

So next, assuming the hyperbola is: $\dfrac{(x-u)^2}{a^2} - \dfrac{(y-v)^2}{b^2} = 1$, we have: $u = 2$, and $v = 0$.

Next, assuming d is the difference between two distances from a point in the hyperbola to the foci, we get: $d = 2a$.   And since the hyperbola has the given point, we can get d the way as follows: $d = \sqrt{(7+3)^2 + (0 - \frac{9}{4})^2} - \frac{9}{4} = \sqrt{\frac{1681}{16}} - \frac{9}{4} = \frac{41}{4} - \frac{9}{4} = 8$.

So we get: $a = 4$.

Next, assuming c is the focal distance, we get: $c = 5$, and $c^2 = a^2 + b^2 \Rightarrow b^2 = c^2 - a^2 \Rightarrow b^2 = 25 - 16 = 9 \Rightarrow b = 3$.

So the hyperbola is: $\dfrac{(x-2)^2}{16} - \dfrac{y^2}{9} = 1$, often put this way, too: $\dfrac{(x-2)^2}{4^2} + \dfrac{y^2}{3^2} = 1$.

If not quite sure of the idea behind the processes above, follow the steps below:

To begin with, if a hyperbola is centered at (u, v), and is horizontal, the equation is: $\dfrac{(x-u)^2}{a^2} - \dfrac{(y-v)^2}{b^2} = 1$, where a is half the transverse axis, and b is half the conjugate.

And if it's <u>horizontal</u>, the center, foci, and vertices share the <u>same y-coordinate</u>. So since the foci given have the same y-coordinates, the hyperbola we want to find is horizontal.

Next, the center is the midpoint between the foci. So the center is: {(0 + 4)/2, (0 + 0)/2}, and thus, is: (2, 0).

Next, in a hyperbola, the difference between two distances from a point to the two foci is constant, and is in fact, the length of the transverse axis.

So assuming d is the difference, we get: $d = 2a$.

And since the given point (-3, 9/4) is in the hyperbola, the difference between the two distances from the given point to the two foci is **2a**, too. And the distance from (-3, 9/4) to (-3, 0) is simply 9/4.   So finding the difference **d**, we get:

$$d = \sqrt{(7+3)^2 + (0 - \tfrac{9}{4})^2} - \tfrac{9}{4} = \sqrt{\tfrac{1681}{16}} - \tfrac{9}{4} = \tfrac{41}{4} - \tfrac{9}{4} = 8.$$

So we get: **a = 4**.   What then, about **b**?

We have a connective equation $c^2 = a^2 + b^2$, where **c** is the focal distance, **2a** is the transverse axis, and **2b** is the conjugate axis. So we get: $b^2 = c^2 - a^2$. What then, about **c**?

The focal distance is the distance from the center to each focus. So since the center is (2, 0), and a focus is (7, 0), we get: **c = 5**.

And thus, we get: $b^2 = c^2 - a^2 = 25 - 16 = 9 \Rightarrow b = 3$.

So the hyperbola is: $\dfrac{(x-2)^2}{16} - \dfrac{y^2}{9} = 1,$ often put this way, too: $\dfrac{(x-2)^2}{4^2} + \dfrac{y^2}{3^2} = 1.$

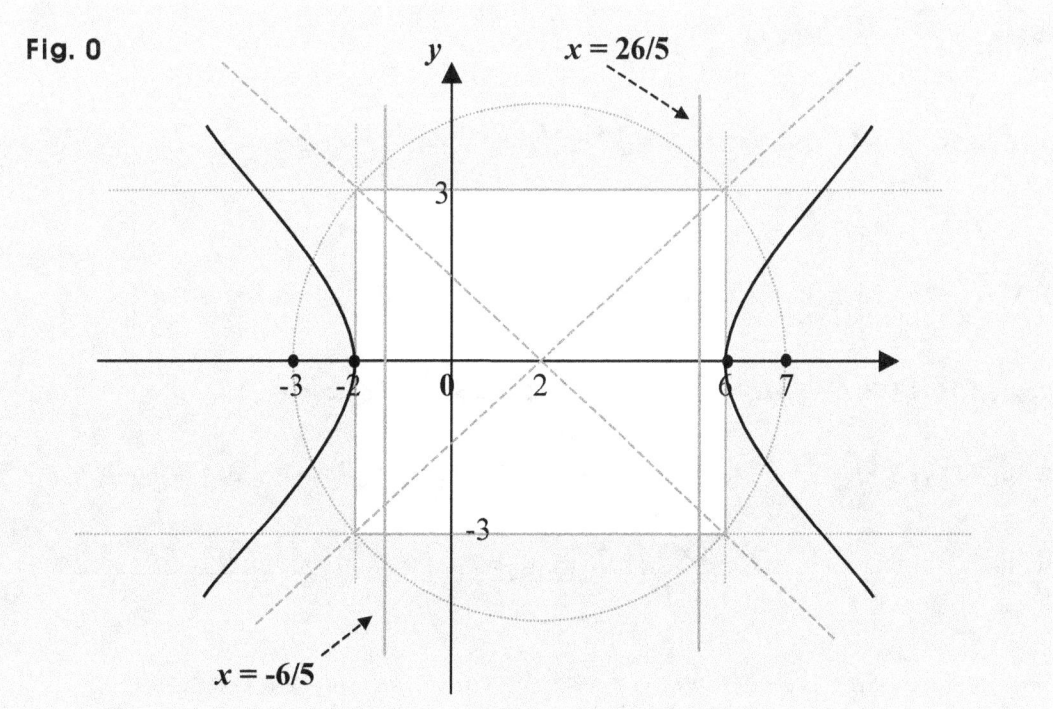

Fig. 0

$x = 26/5$

$x = -6/5$

Suggestions or Solutions
To the Problem in the Example 2

Find the hyperbola that has a point at $(1, 3 - 2\sqrt{2})$, and has the vertices at (4, 5) and (4, 1).

To begin with, the hyperbola is vertical, and the center is (4, 3).

So next, assuming the hyperbola is: $\dfrac{(y-v)^2}{b^2} - \dfrac{(x-u)^2}{a^2} = 1$, we have: $u = 4$, and $v = 1$.

Next, assuming d is the difference between two distances from a point in the hyperbola to the foci, we get: $s = 2a$.   So we get: $2a = 4 \Rightarrow a = 2$.

Next, if c is the focal distance, the foci are $(4, 3 + c)$ and $(4, 3 - c)$.
So next, since the hyperbola has the given point, we can get d the way as follows:

$$d = \sqrt{(1-4)^2 + (3 - 2\sqrt{2} - 3 - c)^2} - \sqrt{(1-4)^2 + (3 - 2\sqrt{2} - 3 + c)^2}$$

$$= \sqrt{9 + (2\sqrt{2} + c)^2} - \sqrt{9 + (2\sqrt{2} - c)^2}$$

$$9 + (2\sqrt{2} + c)^2 = 9 + 8 + 4c\sqrt{2} + c^2 = 17 + 4\sqrt{2}c + c^2$$

$$9 + (2\sqrt{2} - c)^2 = 9 + 8 - 4c\sqrt{2} + c^2 = 17 - 4\sqrt{2}c + c^2$$

And we know: $d = 4$.   So setting: $s = 17 + c^2$, and $t = 4\sqrt{2}c$, we get:

$$\sqrt{s+t} - \sqrt{s-t} = 4 \Rightarrow \sqrt{s+t} = 4 + \sqrt{s-t} \Rightarrow s + t = 16 + 8\sqrt{s-t} + s - t$$

$$\Rightarrow 8\sqrt{s-t} = 2t - 16 \Rightarrow 4\sqrt{s-t} = t - 8 \Rightarrow 16(s - t) = t^2 - 16t + 64 \Rightarrow 16s = t^2 + 64.$$

So we get: $16(17 + c^2) = 32c^2 + 64 \Rightarrow 16c^2 = 208 \Rightarrow c^2 = 13 \Rightarrow c = \sqrt{13}$.

So next, we get: $c^2 = a^2 + b^2 \Rightarrow b^2 = c^2 - a^2 \Rightarrow b^2 = 13 - 4 = 9 \Rightarrow b = 3$.

So the hyperbola is: $\dfrac{(y-3)^2}{9} - \dfrac{(x-4)^2}{16} = 1$, often put this way: $\dfrac{(y-3)^2}{3^2} + \dfrac{(x-4)^2}{4^2} = 1$.

If not quite sure of the idea behind the processes above, follow the steps below:

To begin with, if a hyperbola is centered at (u, v), and is vertical, the equation is:
$\dfrac{(y-v)^2}{b^2} - \dfrac{(x-u)^2}{a^2} = 1$, where b is half the transverse axis, and a is half the conjugate.

And if it's <u>vertical</u>, the center, foci, and vertices share the <u>same x-coordinate</u>. So since the vertices given have the same x-coordinates, the hyperbola we want to find is vertical.

Next, the center is the midpoint between the vertices.
So the center is: $\{(4+4)/2, (5+1)/2\}$, and thus, is: $(4, 3)$. So we get: $u = 4$, and $v = 3$.

Next, in a hyperbola, the difference between two distances from a point to the two foci is constant, and is in fact, the length of the transverse axis.

So assuming d is the difference between the two distances, we get: $d = 2a$, which is the distance between the vertices. And the distance is 4. So we get: $2a = 4 \Rightarrow a = 2$.
What then, about b?

We have a connective equation $c^2 = a^2 + b^2$, where c is the focal distance, $2a$ is the transverse axis, and $2b$ is the conjugate axis. So we get: $b^2 = c^2 - a^2$. What then, about c?

Since c is the focal distance, and the center is $(4, 3)$, the foci are $(4, 3 + c)$ and $(4, 3 - c)$.
So next, since the hyperbola has the given point $(1, 3 - 2\sqrt{2})$, we can put d the way as follows:

$$d = \sqrt{(1-4)^2 + (3 - 2\sqrt{2} - 3 - c)^2} - \sqrt{(1-4)^2 + (3 - 2\sqrt{2} - 3 + c)^2}$$
$$= \sqrt{9 + (2\sqrt{2} + c)^2} - \sqrt{9 + (2\sqrt{2} - c)^2}$$

$$9 + (2\sqrt{2} + c)^2 = 9 + 8 + 4c\sqrt{2} + c^2 = 17 + 4\sqrt{2}c + c^2$$
$$9 + (2\sqrt{2} - c)^2 = 9 + 8 - 4c\sqrt{2} + c^2 = 17 - 4\sqrt{2}c + c^2$$

And we know: $d = 4$. So setting: $s = 17 + c^2$, and $t = 4\sqrt{2}c$, we get:

$$\sqrt{s+t} - \sqrt{s-t} = 4 \Rightarrow \sqrt{s+t} = 4 + \sqrt{s-t} \Rightarrow s+t = 16 + 8\sqrt{s-t} + s - t$$
$$\Rightarrow 8\sqrt{s-t} = 2t - 16 \Rightarrow 4\sqrt{s-t} = t - 8 \Rightarrow 16(s-t) = t^2 - 16t + 64 \Rightarrow 16s = t^2 + 64.$$

So we get: $16(17 + c^2) = 32c^2 + 64 \Rightarrow 16c^2 = 208 \Rightarrow c^2 = 13 \Rightarrow c = \sqrt{13}$.

So next, we get: $c^2 = a^2 + b^2 \Rightarrow b^2 = c^2 - a^2 \Rightarrow b^2 = 13 - 4 = 9 \Rightarrow b = 3$.

So the hyperbola is: $\dfrac{(y-3)^2}{9} - \dfrac{(x-4)^2}{16} = 1$, often put this way: $\dfrac{(y-3)^2}{3^2} + \dfrac{(x-4)^2}{4^2} = 1$.

Fig. 0

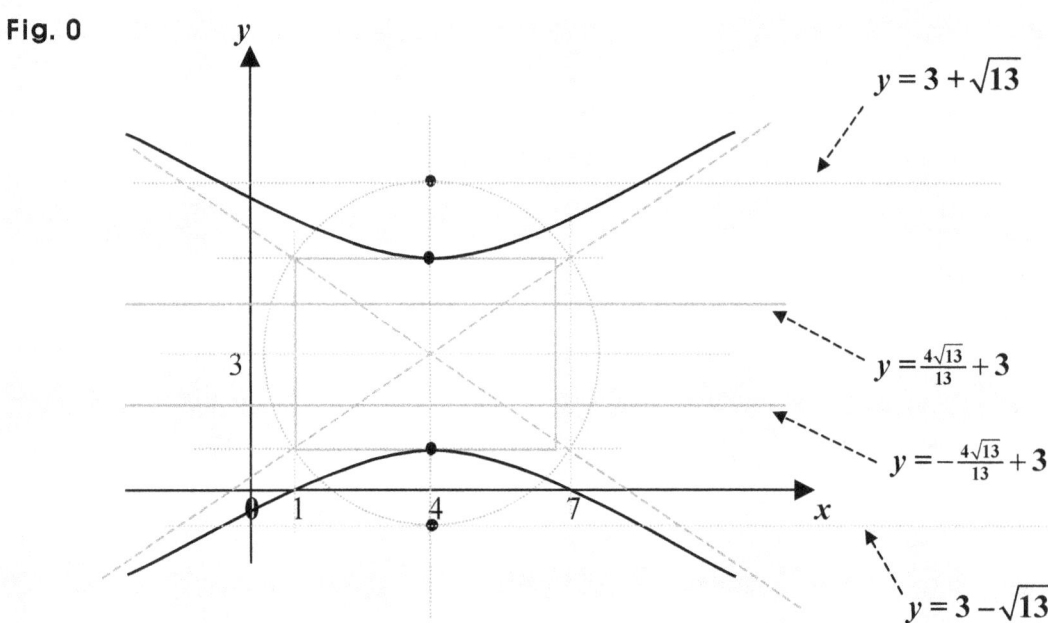

$y = 3 + \sqrt{13}$

$y = \frac{4\sqrt{13}}{13} + 3$

$y = -\frac{4\sqrt{13}}{13} + 3$

$y = 3 - \sqrt{13}$

Suggestions or Solutions
To the Problem in the Example 3

Find the hyperbola of which the eccentricity is 5/3 and a directrix is a line $x = 14/5$, which is corresponding a focus at (6, 2).

Suppose first, $T(x, y)$ is an arbitrary point of a hyperbola, D is the distance from T to a directrix, F is the distance from T to the focus corresponding to the directrix, and e is the eccentricity. Then, we get: $F = eD$, that is, $e = F/D$. So in this case, we get: $F/D = 5/3$.

So next, taking the distances, we get: $D^2 = (x - 14/5)^2$, and $F^2 = (x - 6)^2 + (y - 2)^2$.

Then, we get: $F/D = 5/3 \Rightarrow (F/D)^2 = F^2/D^2 = 25/9 = \dfrac{(x-6)^2 + (y-2)^2}{(x - \frac{14}{5})^2}$

$\Rightarrow 25(x - 14/5)^2 = 9\{(x - 6)^2 + (y - 2)^2\} \Rightarrow \underline{9(x - 6)^2 - 25(x - 14/5)^2} + 9(y - 2)^2 = 0$.

And we have an identity: $A - B = (A + B)(A - B)$.   So we get:

$\underline{9(x - 6)^2 - 25(x - 14/5)^2} = \{3(x - 6) + 5(x - 14/5)\}\{3(x - 6) - 5(x - 14/5)\}$

$= (3x - 18 + 5x - 14)(3x - 18 - 5x + 14) = (8x - 32)(-2x - 4) = -16(x - 4)(x + 2)$

$= -16(x^2 - 2x - 8) = -16(x^2 - 2x + 1 - 9) = -16(x - 1)^2 + 16 \cdot 9 = -16(x - 1)^2 + 12^2$.

So we get: $-16(x - 1)^2 + 12^2 + 9(y - 2)^2 = 0 \Rightarrow 16(x - 1)^2 - 9(y - 2)^2 = 12^2$

$\Rightarrow 4^2(x - 1)^2 - 3^2(y - 2)^2 = 3^2 4^2 \Rightarrow \dfrac{(x-1)^2}{3^2} - \dfrac{(y-2)^2}{4^2} = 1,$ which is the hyperbola we want.

Examples 7 in Hyperbolas

In each example below, the hyperbola is in the *x-y* plane.

0.   Suppose A is a circle centered at (-2, 2), and B is a circle centered at (5, 2). The two circles are away from each other. Suppose now that the radii of A and B increase at the same rate. Then, the two circles meet each other for the first time at a line $x = 1$. In other words, the line $x = 1$ is the line tangent to both circles at the same point at the same time. Suppose further that the radii keep increasing at the same rate thereafter. Then, the two circles A and B keep meeting at two points, and the two points are in a curve, since the two circles are growing. Find the curve.

1.   In the *x-y* plane, find the area where the points satisfy the inequality as follows:
$(x^2 + y^2 - 9)(4x^2 - y^2 - 16) \le 0.$

2.   Assuming $9x^2 + ax - y^2 + by + c = 0$ is a hyperbola, find the values of a, b, and c so that the hyperbola is tangent to the *x*-axis at (1, 0), and passes through (3, 3).

Suggestions or Solutions
To the Problem in the Example 0

Suppose A is a circle centered at (-2, 2), and B is a circle centered at (5, 2). The two circles are away from each other. Suppose now that the radii of A and B increase at the same rate. Then, the two circles meet each other for the first time at a line $x = 1$. In other words, the line $x = 1$ is the line tangent to both circles at the same point at the same time. Suppose further that the radii keep increasing at the same rate thereafter. Then, the two circles A and B keep meeting at two points, and the two points are in a curve, since the two circles are growing. Find the curve.

To begin with, putting the problem in a graph, we can put it the way below:

Fig. 0

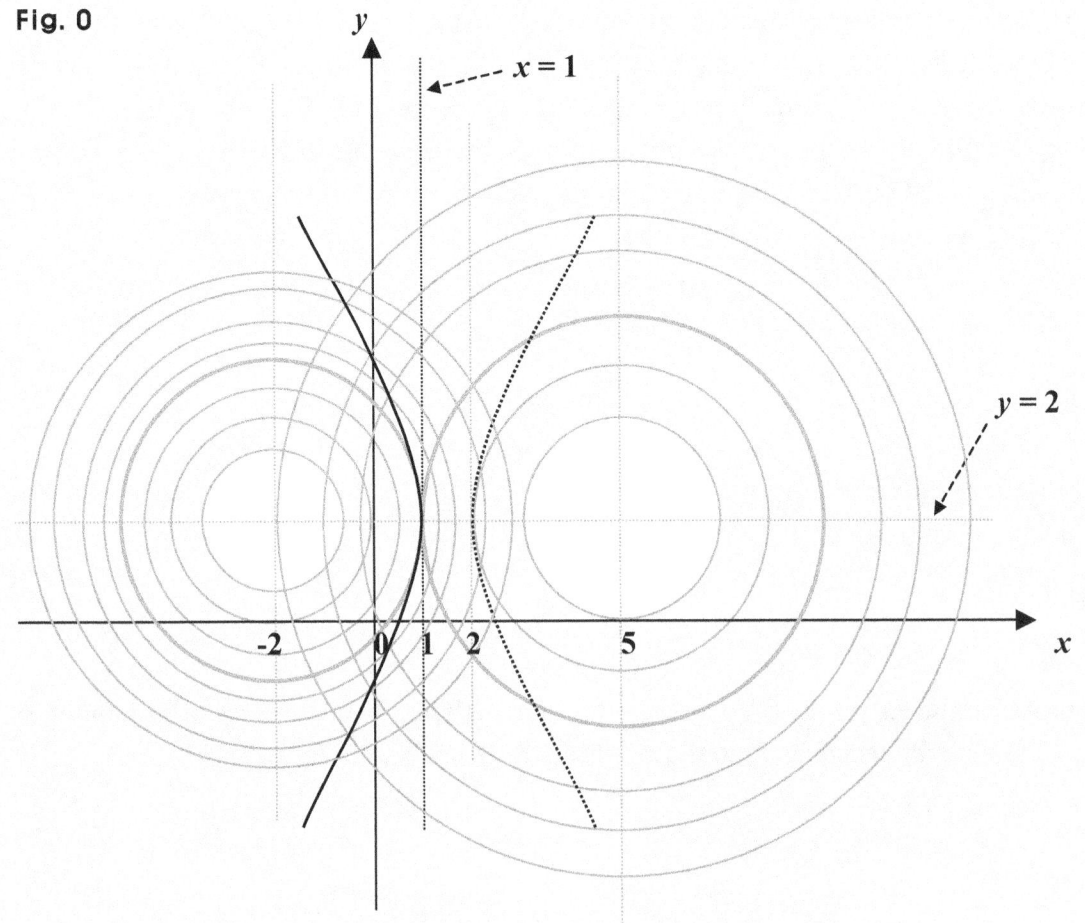

Then, we can see that the curve seems to be a hyperbola.

We want to make sure though, it is the case, and more importantly, we want to get the equation of the curve.

The two radii are growing at the same rate. So the difference between the two radii is constant. So assuming P is one of the two points where the two circles meet, we can say that the difference between the two distances form P to the centers of the two circles is constant. So we can say that the centers of the two circles are the foci of a hyperbola.

Next, the center of a hyperbola is the midpoint between the foci. So since the foci are the centers of the two circles, the center of the hyperbola is: $\{(-2 + 5)/2, (2 + 2)/2\} = (3/2, 2)$.

So since the hyperbola is horizontal, we can put, for now, the equation of the hyperbola the way as follows: $\dfrac{(x-\frac{3}{2})^2}{a^2} - \dfrac{(y-2)^2}{b^2} = 1$. How then, can we get a and b?

The value of a is half the transverse axis, and is the distance from a vertex to the center of the hyperbola. And in this case, one of the two vertices is $(1, 2)$, which is the vertex of the left branch, and is the point where the two circles meet for the first time.

So since the center of the hyperbola is $(3/2, 2)$, the distance from the vertex to the center is $1/2$. And thus, we get: $a = 1/2$.   What then, about b?

Assuming c is the focal distance, we can set: $c^2 = a^2 + b^2$. And we have: $a = 1/2$. And the focal distance is the distance from a focus to the center of the hyperbola. So since $(-2, 2)$ is one of the two foci, and the center is $(3/2, 2)$, we get: $c = 7/2$.

So we get: $c^2 = a^2 + b^2 \Rightarrow b^2 = c^2 - a^2 = 49/4 - 1/4 = 48/4 = 12 \Rightarrow b = 2\sqrt{3}$.

And thus, the hyperbola is: $\dfrac{(x-\frac{3}{2})^2}{(\frac{1}{2})^2} - \dfrac{(y-2)^2}{(2\sqrt{3})^2} = 1$, which can be put the way below, too:

$$4(x-\tfrac{3}{2})^2 - \dfrac{(y-2)^2}{12} = 1.$$

Suggestions or Solutions
To the Problem in the Example 1

In the **x-y** plane, find the area where the points satisfy the inequality as follows:
$(x^2 + y^2 - 9)(4x^2 - y^2 - 16) \leq 0$.

In the graph below, the area we want is the area with slash marks, and the area includes the circle and hyperbola themselves, too.

Fig. 0

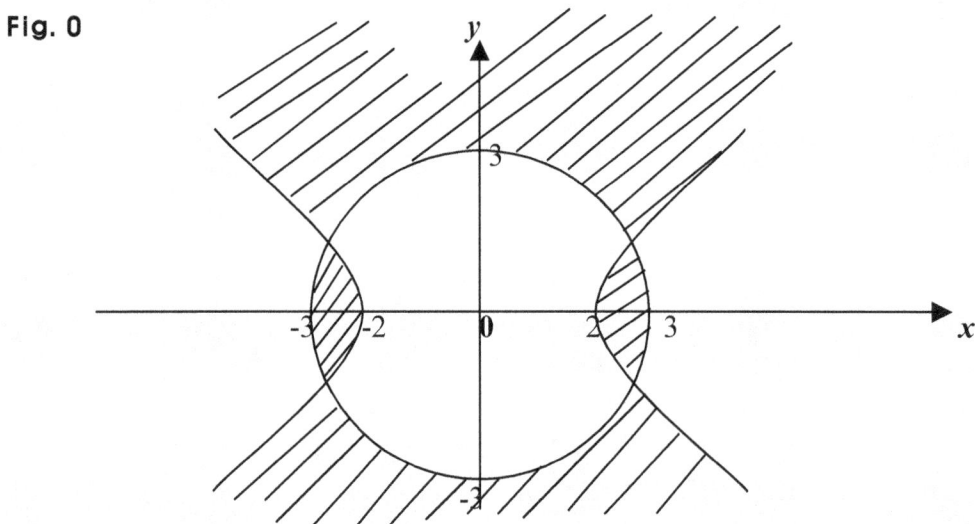

If not quite sure of the idea behind the processes above, follow the steps below:

Assuming first, $AB \leq 0$, we get two cases.
One is: $A \leq 0$ and $B \geq 0$, and the other is: $A \geq 0$ and $B \leq 0$.

So if $(x^2 + y^2 - 9)(4x^2 - y^2 - 16) \leq 0$, we get two cases, too.

One is: $x^2 + y^2 - 9 \leq 0$ and $4x^2 - y^2 - 16 \geq 0$.
And the other is: $x^2 + y^2 - 9 \geq 0$, and $4x^2 - y^2 - 16 \leq 0$.

And we have: $x^2 + y^2 \leq 9 = 3^2$, and $4x^2 - y^2 \geq 16 \Rightarrow x^2/4 - y^2/16 \geq 1 \Rightarrow x^2/2^2 - y^2/4^2 \geq 1$.

So in one case, we have: $x^2 + y^2 \leq 3^2$, and $x^2/2^2 - y^2/4^2 \geq 1$.

And in the other case, we have: $x^2 + y^2 \geq 3^2$, and $x^2/2^2 - y^2/4^2 \leq 1$.

Next, $x^2 + y^2 = 3^2$ is the equation of a circle centered at the origin with a radius of 3.

And $x^2/2^2 - y^2/4^2 = 1$ is the equation of a horizontal hyperbola centered at the origin with a transverse axis of 4 and a conjugate axis of 8.

Next, if $(x - u)^2 + (y - v)^2 \geq r^2$, the inequality means the set of all the points in the area outside a circle centered at (u, v) with a radius of r, and also, the area includes the circle itself, too.

So if $(x - u)^2 + (y - v)^2 < r^2$, the inequality means the set of all the points in only the area inside the circle, so the area does not include the circle itself.

And the same is true for a hyperbola, too.

So if $(x - u)^2/a^2 - (y - v)^2/b^2 < 1$, the inequality means the set of all the points in only the area insides the hyperbola, so the area does not include the hyperbola itself.
The area inside a hyperbola means the area between the braches.

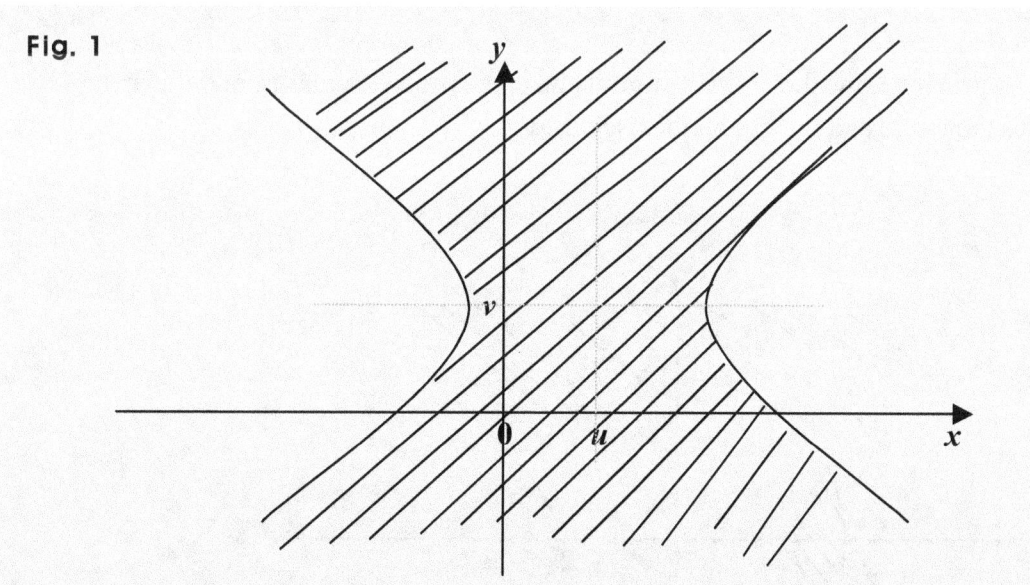

Fig. 1

And if $(x - u)^2/a^2 - (y - v)^2/b^2 \geq 1$, the inequality means the set of all the points in the area outside a hyperbola centered at (u, v) with the main axes of $2a$ and $2b$, and also, the area includes the hyperbola itself, too.

Fig. 2

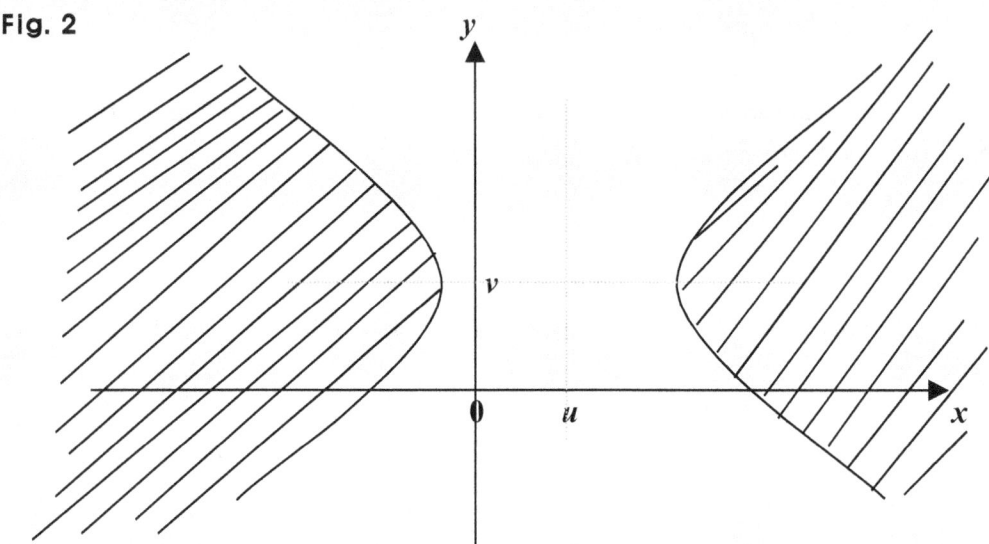

Now, we have two cases.

One is: $x^2 + y^2 \leq 3^2$, and $x^2/2^2 - y^2/4^2 \geq 1$.

And the other is: $x^2 + y^2 \geq 3^2$, and $x^2/2^2 - y^2/4^2 \leq 1$.

So in the graph below, the area we want is the area with slash marks, and the area includes the circle and hyperbola themselves, too.

Fig. 3

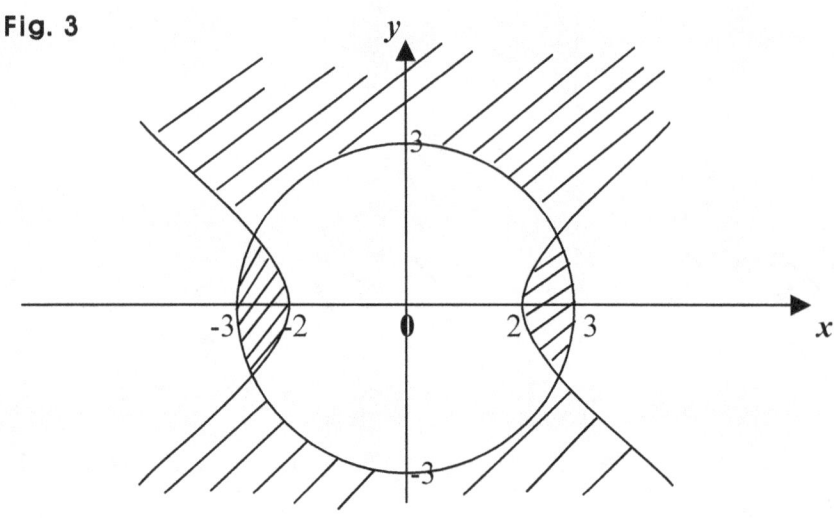

What if we have: $(x^2 + y^2 - 9)(x^2 + 4y^2 - 16) > 0$?

Then, we get two cases, too, but the area is the opposite of the area with slash marks above.

So one is: $x^2 + y^2 > 3^2$, and $x^2/2^2 - y^2/4^2 > 1$.

And the other is: $x^2 + y^2 < 3^2$, and $x^2/2^2 - y^2/4^2 < 1$.

So in the graph above, the area is the area without slash marks, and the area does not include the circle and hyperbola themselves.

In other words, the area is the area with slash marks in the graph below, and the area does not include the circle and hyperbola themselves.

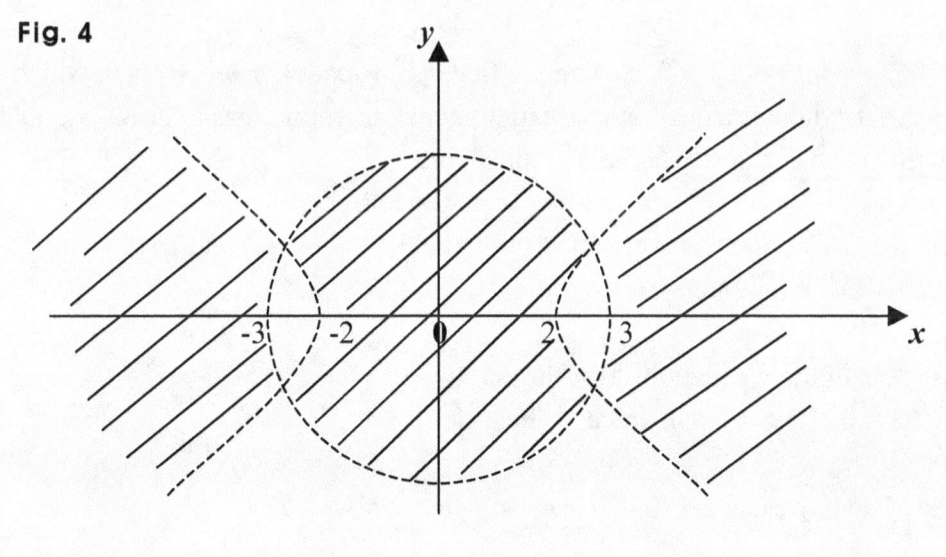

Fig. 4

Suggestions or Solutions
To the Problem in the Example 2

Assuming $9x^2 + ax - y^2 + by + c = 0$ is a hyperbola, find the values of a, b, and c so that the hyperbola is tangent to the x-axis at $(1, 0)$, and passes through $(3, 3)$.

First, if **H** is the hyperbola, since **H** has the point $(3, 3)$, we can get:
$9(3^2) + 3a - 3^2 + 3b + c = 0 \Rightarrow 3a + 3b + c + 72 = 0$.

Next, since **H** is tangent to the x-axis at $(1, 0)$, it has $(1, 0)$, too, so we get: $9 + a + c = 0$.

And next, since the x-axis is tangent to **H**, we can say that a line $y = 0$ meets the hyperbola **H** at one point.
And assuming we find the point where the line meets the hyperbola, we get to solve a system of two equations, one is: $y = 0$, and the other is: $9x^2 + ax - y^2 + by + c = 0$.

And solving the system, since $y = 0$, we can get first, $9x^2 + ax - 0^2 + b \cdot 0 + c = 0$, which is: $9x^2 + ax + c = 0$, which is a quadratic equation, which therefore, has to have a double root, because the line meets the hyperbola at one point.

So the discriminant of the quadratic equation has to be 0.
Thus, we get $a^2 - 4 \cdot 9c = a^2 - 36c = 0$.

So we get a system of three equations as follows:
$3a + 3b + c + 72 = 0$, $9 + a + c = 0$, and $a^2 - 36c = 0$.

So solving the system, we can get first: $9 + a + c = 0 \Rightarrow c = -a - 9$.

So next, we can get: $a^2 - 36c = 0 \Rightarrow a^2 - 36(-a - 9) = 0 \Rightarrow a^2 + 36a + 36 \cdot 9 = 0$.

Meanwhile, $36 \cdot 9 = 4 \cdot 9 \cdot 9 = 2 \cdot 9 \cdot 2 \cdot 9 = 18^2$, and $36 = 2 \cdot 18$.

So we get: $a^2 + 36a + 36 \cdot 9 = a^2 + 2 \cdot 18a + 18^2 = (a + 18)^2 = 0 \Rightarrow a = -18$.

Thus next, we can get: $c = -a - 9 = 18 - 9 = 9 \Rightarrow c = 9$.
And next, $3a + 3b + c + 72 = 0 \Rightarrow 3b = -3a - c - 72 = 54 - 9 - 72 = -27 \Rightarrow b = -9$.

What hyperbola then, is it?

Unlike the case of ellipses, there can be only one hyperbola that can be tangent to the *x*-axis at (1, 0), and pass through (3, 3). And the schematic drawing is as follows:

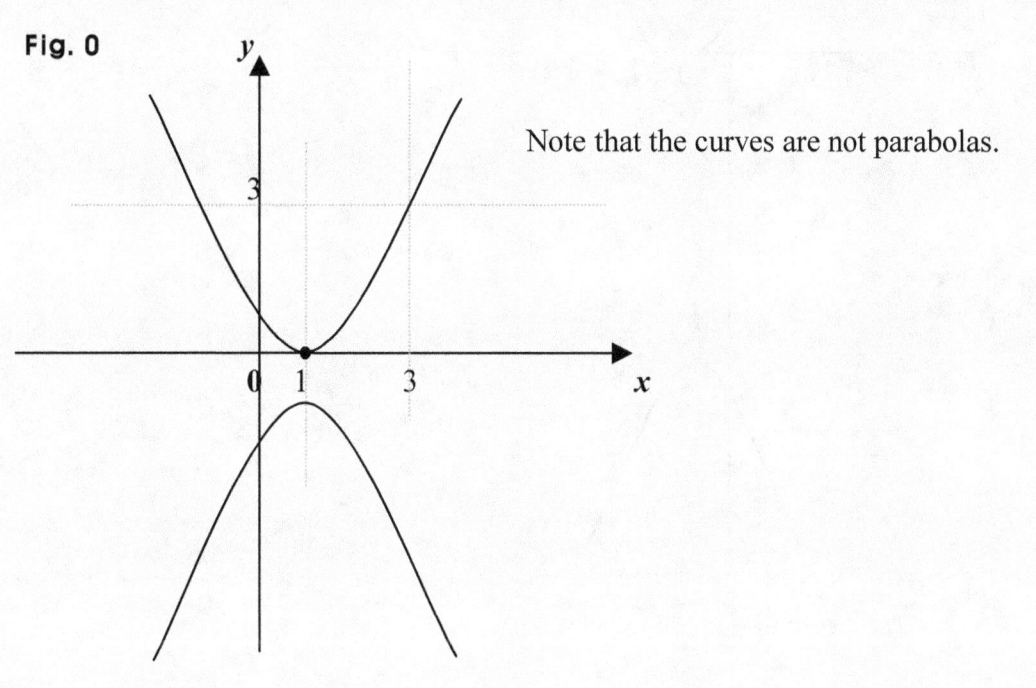

Fig. 0

Note that the curves are not parabolas.

Next, putting values of *a*, *b*, and *c* into the equation $9x^2 + ax - y^2 + by + c = 0$, we get:

$$9x^2 - 18x - y^2 - 9y + 9 = 9(x^2 - 2x + 1 - 1) - \{y^2 + 9y + (\tfrac{9}{2})^2 - (\tfrac{9}{2})^2\} + 9$$

$$= 9(x - 1)^2 - 9 - (y + \tfrac{9}{2})^2 + (\tfrac{9}{2})^2 + 9 = 9(x - 1)^2 - (y + \tfrac{9}{2})^2 + (\tfrac{9}{2})^2 = 0$$

$$\Rightarrow \frac{9(x-1)^2}{(\tfrac{9}{2})^2} - \frac{(y+\tfrac{9}{2})^2}{(\tfrac{9}{2})^2} + 1 = 0 \Rightarrow \frac{(y+\tfrac{9}{2})^2}{(\tfrac{9}{2})^2} - \frac{(x-1)^2}{(\tfrac{3}{2})^2} = 1.$$

So the hyperbola *H* is a vertical hyperbola centered at $(1, -\tfrac{9}{2})$ with the transverse axes of 9 and the conjugate axis of 3. And we can put it in a graph the way below:

Fig. 1

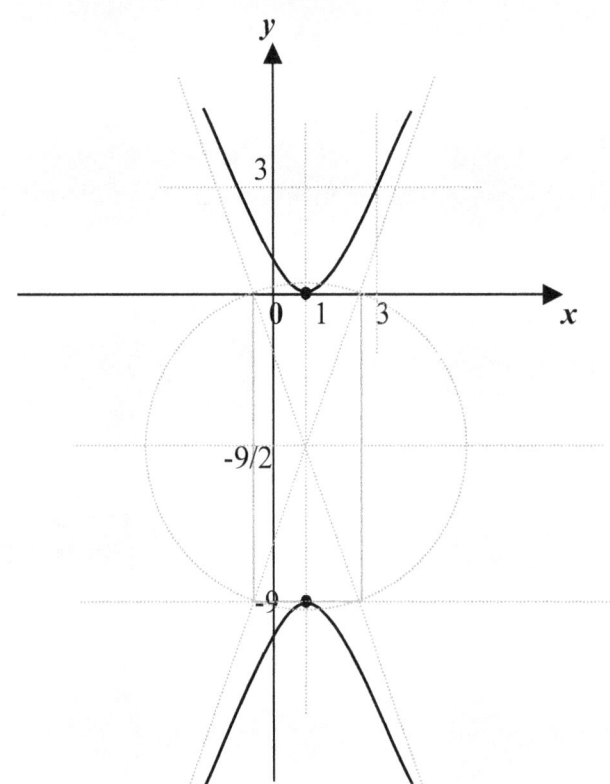

Examples 8 in Hyperbolas

0.   Find the equation of the line tangent to a hyperbola $x^2/a^2 - y^2/b^2 = 1$ at a point (s, t).

.

1.   Suppose that P is a point in a hyperbola, and that A and B are the two foci of the hyperbola. Suppose also, T is the line tangent to the hyperbola at P, M is the angle between the line segment PA and the line T, N is the angle between the line segment PB and the line T, and M and N both are acute angles. Show that $M = N$.

Suggestions or Solutions
To the Problem in the Example 0

Find the equation of the line tangent to a hyperbola $x^2/a^2 - y^2/b^2 = 1$ at a point (s, t).

Assuming first, the tangent line is $y = mx + n$, we get first, $x^2/a^2 - (mx + n)^2/b^2 = 1$. And rearranging terms in the equation above, we get:

$$x^2/a^2 - (mx + n)^2/b^2 = 1 \Rightarrow b^2x^2 - a^2(mx + n)^2 = a^2b^2$$

$$\Rightarrow b^2x^2 - a^2(mx + n)^2 = b^2x^2 - a^2(m^2x^2 + 2mnx + n^2)$$

$$= (b^2 - a^2m^2)x^2 - 2mna^2x - a^2n^2 = a^2b^2 \Rightarrow (b^2 - a^2m^2)x^2 - 2mna^2x - a^2(n^2 + b^2) = 0.$$

Next, the line meets the hyperbola at one point, so the equation's discriminant is 0.

Thus, we get: $m^2n^2a^4 + (b^2 - a^2m^2)a^2(n^2 + b^2) = 0 \Rightarrow m^2n^2a^2 + (b^2 - a^2m^2)(n^2 + b^2) = 0$

$$\Rightarrow b^4 + b^2n^2 - a^2b^2m^2 = 0 \Rightarrow b^2 + n^2 - a^2m^2 = 0.$$

Next, we know the fact that the line passes through (s, t).   So we get: $t = ms + n$. Thus, we get: $n = t - ms$.   So we get:

$$b^2 + n^2 - a^2m^2 = b^2 + (t - ms)^2 - a^2m^2 = 0$$

$$\Rightarrow b^2 + t^2 - 2stm + s^2m^2 - a^2m^2 = m^2(s^2 - a^2) - 2stm + b^2 + t^2 = 0.$$

So we get: $m = \dfrac{st \pm \sqrt{s^2t^2 - (s^2 - a^2)(b^2 + t^2)}}{s^2 - a^2}$.   We know however, m can have one

value only.   So what's inside the square root has to be 0.   That is to say that the discriminant is 0.

So we get: $m = \dfrac{st}{s^2 - a^2}$.   And since the discriminant is 0, we get:

$$s^2t^2 - (s^2 - a^2)(b^2 + t^2) = a^2b^2 + a^2t^2 - s^2b^2 = 0 \Rightarrow b^2(a^2 - s^2) = -a^2t^2 \Rightarrow s^2 - a^2 = \dfrac{a^2t^2}{b^2}.$$

So we get: $m = \dfrac{sb^2}{ta^2}$.   And we know that the line passes through (s, t).

So the tangent line is: $y - t = \dfrac{sb^2}{ta^2}(x - s)$.   And it can be put this way, too: $\dfrac{sx}{a^2} - \dfrac{ty}{b^2} = 1$.

If not quite sure of the idea behind the processes above, follow the steps below:

Suppose first, the tangent line is $y = mx + n$, and is called T, and the hyperbola is H.

Then, since the line T is tangent to the hyperbola H, we can say that the line T meets the hyperbola H at one point.

And assuming we find the point where the line meets the hyperbola, we get to solve a system of two equations, one is $y = mx + n$, and the other is $x^2/a^2 - y^2/b^2 = 1$.

And solving the system, since $y = mx + n$, we can get first, $x^2/a^2 - (mx + n)^2/b^2 = 1$, which is: $b^2x^2 - a^2(mx + n)^2 = a^2b^2$, which is a quadratic equation, which therefore, has to have a double root, because the line meets the hyperbola at one point.

So the discriminant of the quadratic equation has to be 0.
By the way, if the coefficient of x is even, we can take a quarter of the discriminant, because the result will get simplified that way.

If for instance, we have: $Ax^2 + 2Bx + C = 0$, the discriminant is: $B^2 - AC$.

So first, rearranging terms in the equation, we get:

$$b^2x^2 - a^2(mx + n)^2 = b^2x^2 - a^2(m^2x^2 + 2mnx + n^2)$$

$$= (b^2 - a^2m^2)x^2 - 2mna^2x - a^2n^2 = a^2b^2 \Rightarrow (b^2 - a^2m^2)x^2 - 2mna^2x - a^2(n^2 + b^2) = 0.$$

So next, getting the quarter of the discriminant, we get:

$$m^2n^2a^4 + (b^2 - a^2m^2)a^2(n^2 + b^2) = 0 \Rightarrow m^2n^2a^2 + (b^2 - a^2m^2)(n^2 + b^2) = 0$$
$$\Rightarrow m^2n^2a^2 + b^2n^2 + b^4 - m^2n^2a^2 - a^2b^2m^2 = b^4 + b^2n^2 - a^2b^2m^2 = 0$$
$$\Rightarrow b^2 + n^2 - a^2m^2 = 0.$$

Next, knowing the slope and a point of a line, we can get the equation of the line.

We know the tangent line T passes through (s, t), and the slope is m.
And also, $s, t, a,$ and b are all known values.

So finding m, we can get the line T.   How then, can we find m?

We now have two equations that have **m**.

One is: $b^2 + n^2 - a^2m^2 = 0$. and the other is: $y = mx + n$.

And we know the fact that **T** passes through **(s, t)**.   So we get: $t = ms + n$.
Thus, we get: $n = t - ms$.

So we can now get **m** putting **n** into the other equation.   Then, we get:

$$b^2 + n^2 - a^2m^2 = b^2 + (t - ms)^2 - a^2m^2 = 0$$
$$\Rightarrow b^2 + t^2 - 2stm + s^2m^2 - a^2m^2 = m^2(s^2 - a^2) - 2stm + b^2 + t^2 = 0, \text{ which is quadratic.}$$

So using the quadratic formula, we get: $m = \dfrac{st \pm \sqrt{s^2t^2 - (s^2 - a^2)(b^2 + t^2)}}{s^2 - a^2}$.

We know however, **m** can have one value only.

So what's inside the square root has to be 0. That is to say that the discriminant is 0.

So we get: $m = \dfrac{st}{s^2 - a^2}$.

And since the discriminant is 0, we get:

$$s^2t^2 - (s^2 - a^2)(b^2 + t^2) = s^2t^2 - s^2b^2 - s^2t^2 + a^2b^2 + a^2t^2 = -s^2b^2 + a^2b^2 + a^2t^2 = 0$$
$$\Rightarrow a^2b^2 - s^2b^2 + a^2t^2 = 0 \Rightarrow b^2(b^2 - s^2) = -a^2t^2 \Rightarrow b^2(s^2 - a^2) = a^2t^2 \Rightarrow s^2 - a^2 = \dfrac{a^2t^2}{b^2}.$$

So we get: $m = \dfrac{st}{s^2 - a^2} = st\dfrac{b^2}{a^2t^2} = \dfrac{sb^2}{ta^2}$, which is the slope of the line **T**.

And we know that the line **T** passes through **(s, t)**.
If a line has a slope of **k**, and has a point **(p, q)**, the line is: $y - q = k(x - p)$.

So the line **T** is: $y - t = \dfrac{sb^2}{ta^2}(x - s)$, which looks a bit complicated.

So simplifying the equation above, we can get:

$$y - t = \frac{sb^2}{ta^2}(x - s) \Rightarrow ta^2(y - t) = sb^2(x - s) \Rightarrow tya^2 - t^2a^2 = sxb^2 - s^2b^2$$

$$\Rightarrow sxb^2 - tya^2 = s^2b^2 - t^2a^2 \Rightarrow sx - \frac{tya^2}{b^2} = s^2 - \frac{t^2a^2}{b^2} \Rightarrow \frac{sx}{a^2} - \frac{ty}{b^2} = \frac{s^2}{a^2} - \frac{t^2}{b^2}.$$

And we know that the hyperbola H passes through the point (s, t), and H is:
$$\frac{x^2}{a^2} - \frac{y^2}{b^2} = 1.$$

So we get: $\dfrac{s^2}{a^2} - \dfrac{t^2}{b^2} = 1,$ and thus, the tangent line is: $\dfrac{sx}{a^2} - \dfrac{ty}{b^2} = 1.$

So for instance, finding the line tangent to $\dfrac{x^2}{3^2} - \dfrac{y^2}{2^2} = 1$ at $(4, \frac{2\sqrt{7}}{3})$, we get:

$$\frac{4x}{3^2} - \frac{2\sqrt{7}y}{3 \cdot 2^2} = \frac{4x}{9} - \frac{\sqrt{7}y}{6} = 1,$$ which is the tangent line.

And of course, we can put it this way, too: $8x - 3\sqrt{7}y - 18 = 0.$

What if we are given the slope of the tangent line and are asked to find the tangent line?

Assuming the slope given is m, we can assume that the tangent line is $y = mx + n$.

Then, finding n, we get the line tangent to the hyperbola.

Finding n, we just solve $b^2 + n^2 - a^2m^2 = 0$ for n. And we get the equation taking the discriminant of $x^2/a^2 - (mx + n)^2/b^2 = 1$, which is: $b^2x^2 - a^2(mx + n)^2 = a^2b^2.$

So solving it for n, we get: $n = \pm\sqrt{a^2m^2 - b^2}.$ And thus, we get two tangent lines.

One is: $y = mx + \sqrt{a^2m^2 - b^2}$, and the other is: $y = mx - \sqrt{a^2m^2 - b^2}$.

Fig. 0

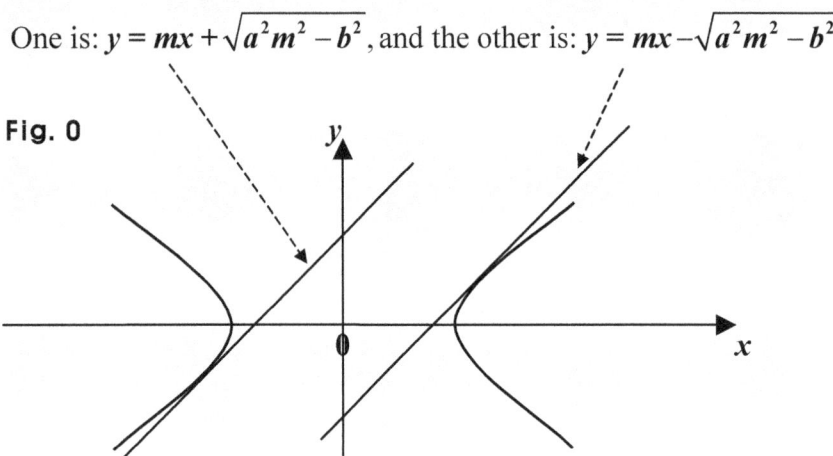

So for instance, finding the lines that have a slope of 2, and are tangent to $\dfrac{x^2}{3^2} - \dfrac{y^2}{2^2} = 1$,

we get: $y = 2x \pm \sqrt{3^2 2^2 - 2^2}$. So we get two tangent lines.

One is $y = 2x + 4\sqrt{2}$, and the other is: $y = 2x - 4\sqrt{2}$.

What if this time, we want to find the line tangent to an hyperbola

$\dfrac{(x-u)^2}{a^2} + \dfrac{(y-v)^2}{b^2} = 1$ at a point (s, t)?

Translating the hyperbola $\dfrac{x^2}{a^2} - \dfrac{y^2}{b^2} = 1$ in the amount of u along the x-axis, and in the

amount of v along the y-axis, we get the hyperbola $\dfrac{(x-u)^2}{a^2} - \dfrac{(y-v)^2}{b^2} = 1$.

So finding the line tangent to $\dfrac{(x-u)^2}{a^2} - \dfrac{(y-v)^2}{b^2} = 1$ at the point (s, t), we find first the

line tangent to $\dfrac{x^2}{a^2} - \dfrac{y^2}{b^2} = 1$ at $(s - u, t - v)$. Then, we get: $\dfrac{(s-u)x}{a^2} - \dfrac{(t-v)y}{b^2} = 1$.

Next, translating the line above in the amount of *u* along the *x*-axis, and in the amount of

v along the *y*-axis, we get the line tangent to $\dfrac{(x-u)^2}{a^2} - \dfrac{(y-v)^2}{b^2} = 1$ at the point *(s, t)*.

Then, we get: $\dfrac{(s-u)(x-u)}{a^2} - \dfrac{(t-v)(y-v)}{b^2} = 1,$ which is the line tangent to

$\dfrac{(x-u)^2}{a^2} - \dfrac{(y-v)^2}{b^2} = 1$ at the point *(s, t)*.

Fig. 1

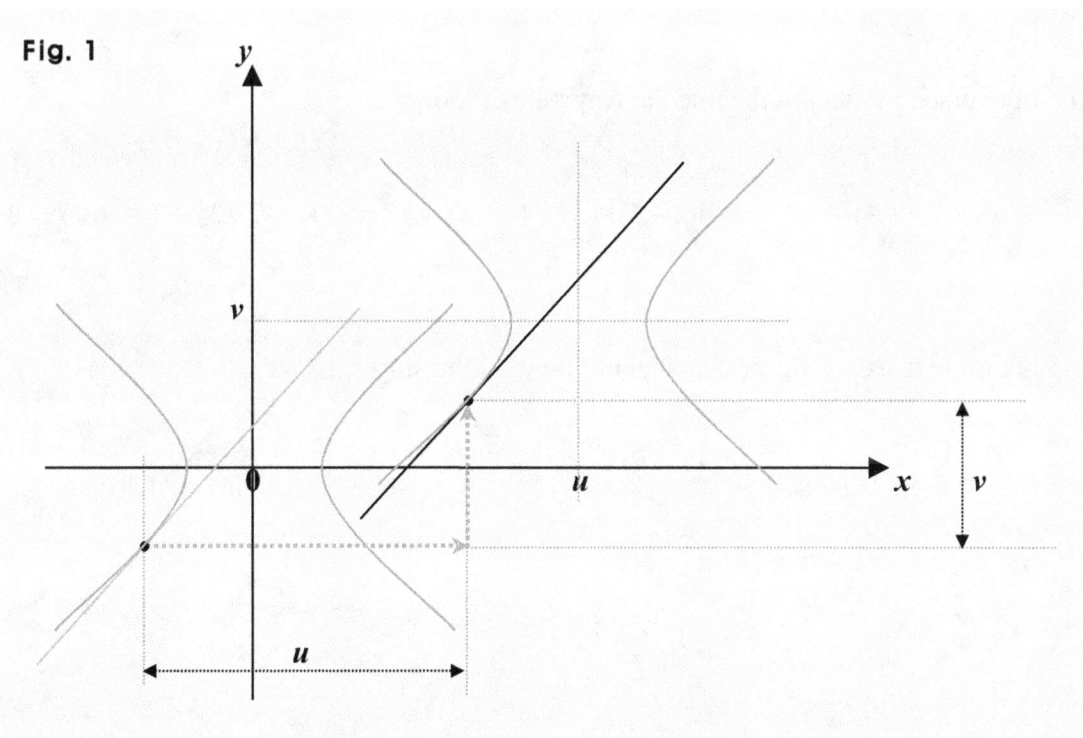

So for instance, finding the line tangent to $\dfrac{(x-1)^2}{3^2} - \dfrac{(y-2)^2}{2^2} = 1$ at $(-1, \frac{2\sqrt{7}}{3}+2)$, we can

get first, the line tangent to $\dfrac{x^2}{3^2} - \dfrac{y^2}{2^2} = 1$ at $(-1-1, \frac{2\sqrt{7}}{3}+2-2)$, that is, $(-2, \frac{2\sqrt{7}}{3})$.

Then, the line is: $\dfrac{-2x}{3^2} + \dfrac{\frac{2\sqrt{7}}{3}\,y}{2^2} = 1.$

Next, translating the line above in the amount of 1 along the **x**-axis, and in the amount of 2 along the **y**-axis, we get the line tangent to $\dfrac{(x-1)^2}{3^2} + \dfrac{(y-2)^2}{2^2} = 1$ at $(-1, \frac{2\sqrt{7}}{3} + 2)$.

Then, the line is: $\dfrac{-2(x-1)}{3^2} - \dfrac{\frac{2\sqrt{7}}{3}(y-2)}{2^2} = 1$, which is $\dfrac{-2(x-1)}{9} - \dfrac{\sqrt{7}(y-2)}{6} = 1$, which is the line tangent to $\dfrac{(x-1)^2}{3^2} - \dfrac{(y-2)^2}{2^2} = 1$ at $(-1, \frac{2\sqrt{7}}{3} + 2)$.

And of course, we can put the line the way below, too:

$$\dfrac{-2(x-1)}{9} - \dfrac{\sqrt{7}(y-2)}{6} = 1 \implies -4(x-1) - 3\sqrt{7}(y-2) = 18 \implies 4x + 3\sqrt{7}y + 14 - 6\sqrt{7} = 0.$$

And the same is true, too, for the tangent line with the slope given.

So the lines that are tangent to $\dfrac{(x-u)^2}{a^2} - \dfrac{(y-v)^2}{b^2} = 1$, and have a slope of **m** are as follows: $y - v = m(x - u) \pm \sqrt{a^2 m^2 - b^2}$.

Suggestions or Solutions
To the Problem in the Example 1

Suppose P is a point in an hyperbola, A and B are the foci, and T is a line tangent to the hyperbola at P.   Suppose also, m and n are two acute angles, m is an angle between PA and T, and n is an angle between PB and T.   Then, show that $m = n$.

For simplicity, let's use a horizontal hyperbola centered at the origin.
And suppose that the hyperbola is H, and the equation is $x^2/a^2 - y^2/b^2 = 1$.
Then, a is half the transverse axis, b is half the conjugate axis, and we get: $c^2 = a^2 + b^2$ where c is the focal distance.
And putting now, the problem description in a graph, we can put it the way below:

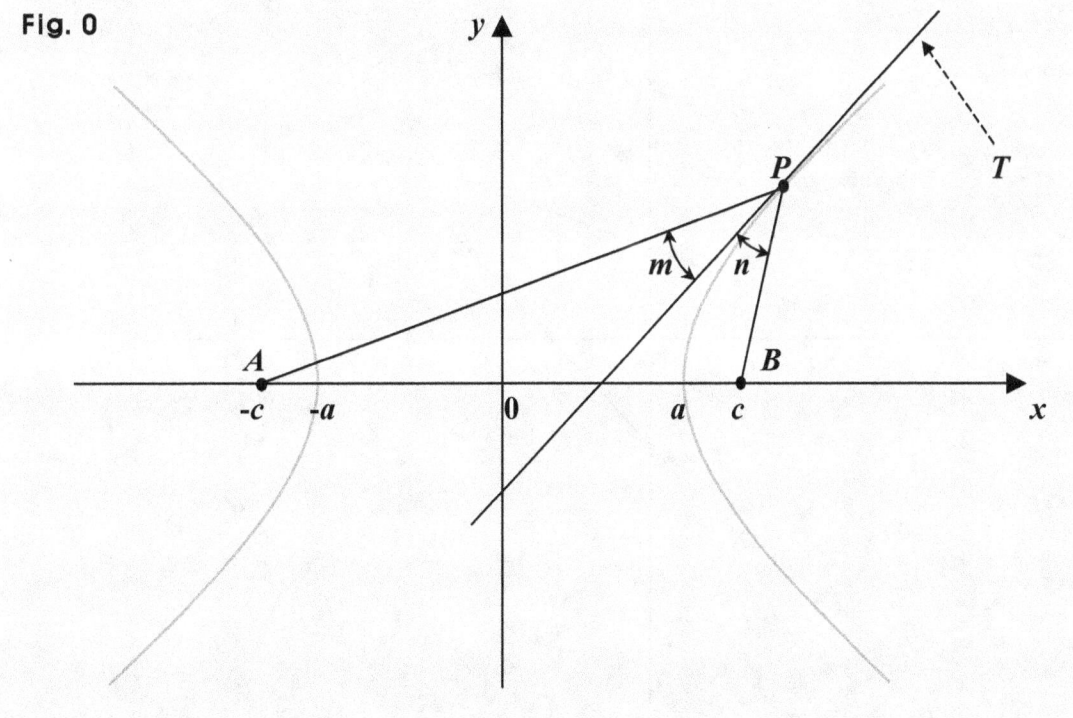

Fig. 0

Next, we have a fact that a line tangent to an ellipse also bisects the angle made by line segments connecting the foci and a point in the ellipse. It is covered in the example 4 in
Examples 5 in Ellipses.

So let's put in a graph, along with the hyperbola **H**, an ellipse that has the same foci as the foci **H** has, and also, passes through the point **P**, too. If an ellipse and a hyperbola have the same foci, the are said to be confocal.

Suppose the ellipse is **E**, and the equation is: $x^2/w^2 + y^2/z^2 = 1$.
Then, **w** is the major radius, **z** is the minor radius, and we get: $c^2 = w^2 - z^2$ where **c** is the focal distance, which is the same as the focal distance of **H**. And putting in the graph, the ellipse, together with some other line and points, we can put them the way below:

Fig. 1

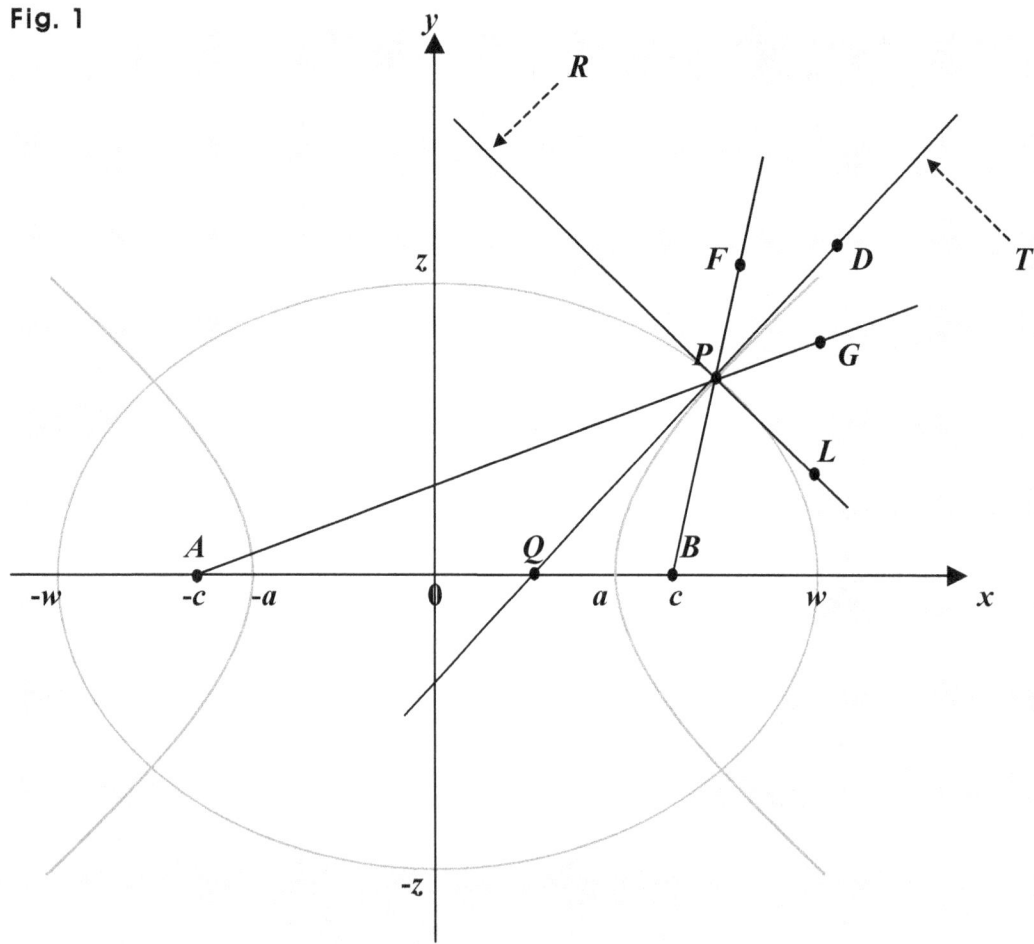

Then, **R** is a line tangent to the ellipse **E** at **P**, and the line **R** is perpendicular to the line **T**. And we will get to see how it is the case shortly.

What does **R** though, have to do with the fact that **T** bisects the angle **APB**?

First, R bisects the angle BPG, so the angle BPL is the same as the angle GPL.

Next, R is perpendicular to T, the angle QPL and the angle DPL are both 90°. So the angle QPB is the same as the angle DPG.

Next, the angle APQ and the angle DPG are opposite angles, so both are the same. So the angle APQ is the same as the angle BPQ.

Therefore, the tangent line T bisects the angle APB.

So showing that R is perpendicular to T, we show that the two angles m and n are equal.

To begin with, we have a fact that if one line is perpendicular to another, the product of the slopes of the two lines is -1.
So showing that the product of the slopes of R and T is -1, we show: $m = n$.

Suppose first, P is at (s, t).

Then, since the hyperbola H is: $x^2/a^2 - y^2/b^2 = 1$, the tangent line T is: $y - t = \dfrac{sb^2}{ta^2}(x - s)$,

found in the example 0 above.   So the slope of T is: $\dfrac{sb^2}{ta^2}$.

And of course, we can put the line T this way, too: $\dfrac{sx}{a^2} - \dfrac{ty}{b^2} = 1$.

Next, the ellipse E is: $x^2/w^2 + y^2/z^2 = 1$.
And we know that the line R is tangent to E at $P(s, t)$, too.

So the line R is: $y - t = -\dfrac{sz^2}{tw^2}(x - s)$, found in the example 2 in **Examples 4 in**

Ellipses.   So the slope of R is: $\dfrac{sz^2}{tw^2}$.

And of course, we can put the line R this way, too: $\dfrac{sx}{w^2} + \dfrac{ty}{z^2} = 1$.

Thus, the product of the slopes of R and T is: $-\dfrac{sz^2}{tw^2} \cdot \dfrac{sb^2}{ta^2} = -\left(\dfrac{bsz}{atw}\right)^2$.

And **H** and **E** both pass through **P(s, t)**.   So we get: $s^2/a^2 - t^2/b^2 = 1$ and $s^2/w^2 + t^2/z^2 = 1$.

And setting next, $u = s^2$ and $v = t^2$, we get:

$$s^2/a^2 - t^2/b^2 = 1 \Rightarrow u/a^2 - v/b^2 = 1 \Rightarrow ub^2 - va^2 = a^2b^2 \ \text{.......(1)}$$

$$s^2/w^2 + t^2/z^2 = 1 \Rightarrow u/w^2 + v/z^2 = 1 \Rightarrow uz^2 + vw^2 = w^2z^2 \ \text{.......(2)}$$

Let' first, eliminate **v**. Then, multiplying (1) by w^2, and multiplying (2) by a^2, we get:

$$w^2(ub^2 - va^2) = w^2(a^2b^2) \Rightarrow ub^2w^2 - va^2w^2 = a^2b^2w^2 \ \text{........(3)}$$

$$a^2(uz^2 + vw^2) = a^2(w^2z^2) \Rightarrow ua^2z^2 + va^2w^2 = w^2z^2a^2 \ \text{.........(4)}$$

And next, taking: (3) + (4), we get:

$$ub^2w^2 + ua^2z^2 = u(b^2w^2 + a^2z^2) = a^2b^2w^2 + w^2z^2a^2.$$

So we get: $u = s^2 = \dfrac{a^2b^2w^2 + w^2z^2a^2}{b^2w^2 + a^2z^2}.$

Next, to eliminate **u**, multiplying (1) by z^2, and multiplying (2) by b^2, we get:

$$ub^2 - va^2 = a^2b^2 \Rightarrow z^2(ub^2 - va^2) = z^2a^2b^2 \Rightarrow ub^2z^2 - va^2z^2 = z^2a^2b^2 \ \text{........(5)}$$

$$uz^2 + vw^2 = w^2z^2 \Rightarrow b^2(uz^2 + vw^2) = b^2w^2z^2 \Rightarrow ub^2z^2 + vb^2w^2 = b^2w^2z^2 \ \text{......(6)}$$

Subtracting (5) from (6) above, we get:

$$vb^2w^2 + va^2z^2 = v(b^2w^2 + a^2z^2) = b^2w^2z^2 - z^2a^2b^2.$$

So we get: $v = t^2 = \dfrac{b^2w^2z^2 - z^2a^2b^2}{b^2w^2 + a^2z^2}.$

Then, we get: $\dfrac{u}{v} = \dfrac{s^2}{t^2} = s^2 \cdot \dfrac{1}{t^2} = \dfrac{a^2 b^2 w^2 + w^2 z^2 a^2}{b^2 w^2 + a^2 z^2} \cdot \dfrac{b^2 w^2 + a^2 z^2}{b^2 w^2 z^2 - z^2 a^2 b^2} = \dfrac{a^2 b^2 w^2 + w^2 z^2 a^2}{b^2 w^2 z^2 - z^2 a^2 b^2}.$

So we get: $\dfrac{s^2}{t^2} = \dfrac{a^2 b^2 w^2 + w^2 z^2 a^2}{b^2 w^2 z^2 - z^2 a^2 b^2}.$

Thus, we get: $\left(\dfrac{bsz}{atw}\right)^2 = \dfrac{s^2}{t^2} \cdot \dfrac{b^2 z^2}{a^2 w^2} = \dfrac{a^2 b^2 w^2 + w^2 z^2 a^2}{b^2 w^2 z^2 - z^2 a^2 b^2} \cdot \dfrac{b^2 z^2}{a^2 w^2} = \dfrac{b^2 z^2 (a^2 b^2 w^2 + w^2 z^2 a^2)}{a^2 w^2 (b^2 w^2 z^2 - z^2 a^2 b^2)}$

$= \dfrac{b^2 z^2 a^2 w^2 (b^2 + z^2)}{a^2 w^2 b^2 z^2 (w^2 - a^2)} = \dfrac{b^2 + z^2}{w^2 - a^2}.$

So the product of the two slopes is: $-\left(\dfrac{bsz}{atw}\right)^2 = -\dfrac{b^2 + z^2}{w^2 - a^2}.$

Next, we have:

$c^2 = a^2 + b^2 \Rightarrow b^2 = c^2 - a^2$ in the case of the ellipse.

$c^2 = w^2 - z^2 \Rightarrow z^2 = w^2 - c^2$ in the case of the hyperbola.

So we get: $b^2 + z^2 = c^2 - a^2 + w^2 - c^2 = w^2 - a^2.$

And thus, the product is: $-\left(\dfrac{bsz}{atw}\right)^2 = -\dfrac{b^2 + z^2}{w^2 - a^2} = -\dfrac{w^2 - a^2}{w^2 - a^2} = -1.$

So we can now say that **R** is perpendicular to **T**, and consequently, we get: **m = n**.

By the way, the two lines **R** and **T** are called normal lines.
A normal line is a line perpendicular to a surface or a curve, is defined at a point in the surface or curve, and is in particular, said to be normal to the surface or the curve at the point.
So **R** is said to be normal to the hyperbola **H** at **P(s, t)**, and **T** is normal to the ellipse **E** at **P(s, t)**.

And let's now get back to the physical property of a hyperbola.

In the figure below, **T** is a line bisecting the angle **APB**, so the angle **APQ** is the same as the angle **FPD**.

And in fact, all the four angles, $\angle APQ$, $\angle QPB$, $\angle FPD$, and $\angle GPD$ are all the same.

Therefore, if a ray emits from the focus **A**, and contacts **P**, it is reflected as if it came from the focus **B**. That is, it gets to proceed in the direction of **PF**.

And if a ray emits from the focus **B**, and contacts **P**, it is reflected as if it came from the focus **A**. That is, it gets to proceed in the direction of **PG**.

Fig. 2

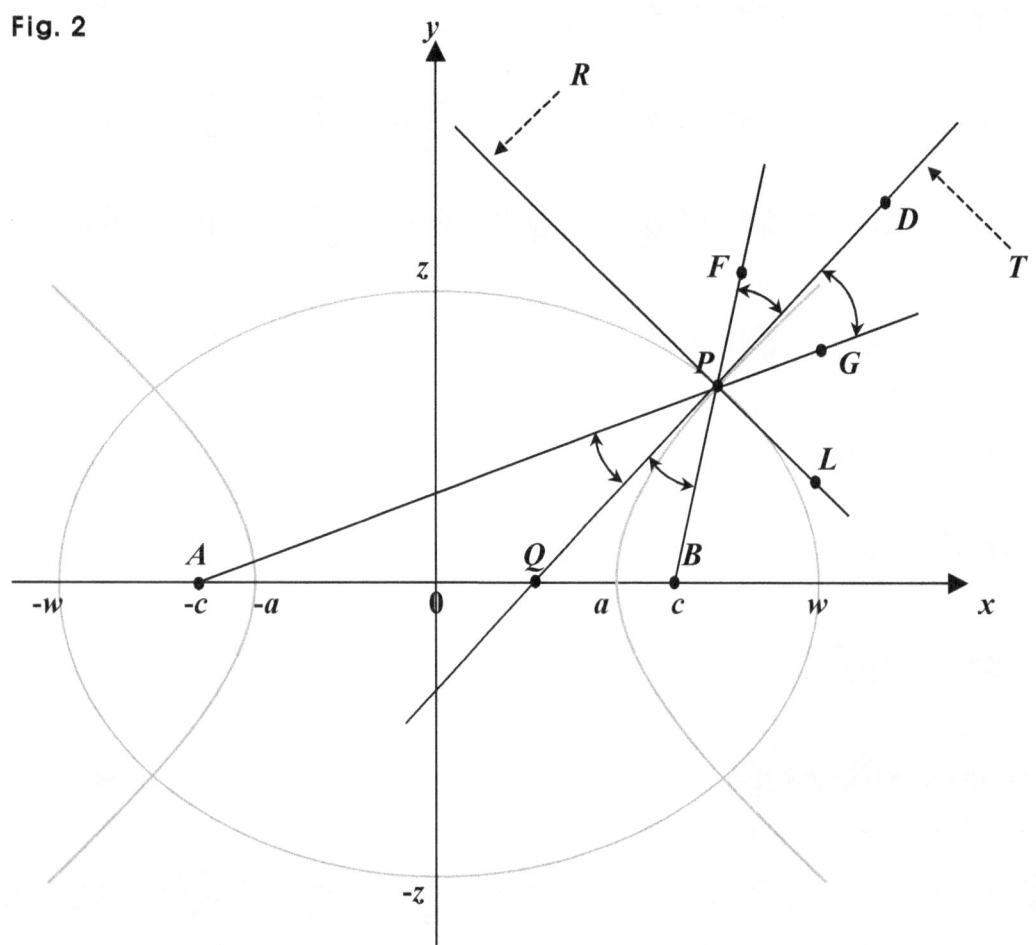